T0146330

The Art of Gunsmithing –
The Shotgun

The Art of Gunsmithing –
The Shotgun

Lewis Potter

The Crowood Press

First published in 2006 by
The Crowood Press Ltd
Ramsbury, Marlborough
Wiltshire SN8 2HR

www.crowood.com

This impression 2021

© Lewis Potter 2006

All rights reserved. No part of this publication may be reproduced or transmitted in
any form or by any means, electronic or mechanical, including photocopy, recording,
or any information storage and retrieval system, without permission in writing from
the publishers.

British Library Cataloguing-in-Publication Data
A catalogue record for this book is available from the British Library.

ISBN 978 1 86126 815 0

Dedication
This book is dedicated to my friend and mentor, G.L. (Geoff) Hart, Gunmaker.

Disclaimer
The author and the publisher do not accept responsibility in any manner whatsoever
for any error or omission, nor any loss, damage, injury, or adverse outcome, or liability
of any kind incurred as a result of the use of the information contained in this book,
or reliance upon it.

Frontispiece: Checking by sight that wired-on ribs are true

Except the photograph on page 141, all illustrations in this book are by the author.

Typeset by Focus Publishing, Sevenoaks, Kent
Printed and bound in India by Parksons Graphics Pvt. Ltd.

Contents

Acknowledgements

I thank The Crowood Press for their help. I can happily blame my friend, George Wallace, retired firearms officer of the BASC, for volunteering me for this project – I know he was, as ever, only doing his best to help another member of the shooting community.

I thank Chris Price, proprietor of Helston Gunsmiths and Master of the Worshipful Company of Gunmakers, for his enthusiasm in volunteering to proof-read most of the book and his helpful advice and comments; Ken Halbert, foreman at Westley Richards, for his help with historical detail; and Peter Spode for his contribution and guidance about engraving. I wish also to thank the following who have helped or assisted in any way, often at short notice: the Proof Master and staff of the Birmingham Proof House; Graham N Greener (W W Greener Sporting Guns Ltd); Mark Crudgington (Crudgington Gunmakers) and Peter Thomson; Pete Green for his computer advice; Pierre Shone for assistance with photography; Nigel Teague (Teague Precision Chokes); Tony White (T R White & Co, Gunmakers); Graham Walker and Lawrence and Yvonne Langridge who always enthusiastically support my various projects. In fact, my appreciation goes to all who have shown their kindness by supporting this project, particularly those who let me photograph their guns for inclusion in this book.

It is not even all my own work. My wife, Sue, using considerable ability and determination, managed as always to turn my handwritten scribble into legible print. My daughter, Lucie Ellen, a student of the English language, exercised her skills by correcting the grammar. When required, my sons, Matt and Dan, proved good male models and Jamie helped whenever he could. Truly a family effort.

Foreword

The gun trade in the United Kingdom is quite rightly regarded as one that boasts some of the best gunmakers in the world, not just from the gun-making centres of London and Birmingham, but also from the provinces. Less famous are the 'best gunsmiths', who look after the shooting sportsman and woman with gun repairs, alterations and refurbishment.

Many gunsmiths, whether self-taught or drawing on experience gained during time served under a skilled gunsmith or gunmaker, like to preserve the mystery of the art of the gunsmith by protecting their secrets. Fortunately, Lewis Potter, a true crafts-man and accomplished gunsmith, is not so selfish. His book, *The Art of Gunsmithing*, is long overdue. Each chapter specializes in an area covered by most competent gunsmiths, describing the problems encountered and some of the tricks of the trade used to effect a satisfactory repair. Two things strike me more than anything else with the content of this book: first, the way it is written makes for compulsive reading; second, and more importantly, it clearly reflects the author's commitment to quality.

Lewis is not prepared to send out a poor-quality job, and does not mince his words when he encounters the work of the 'bodger'.

Those of us who have spent a lifetime in the gun trade enjoy the company of fellow craftsmen -- there is so much to talk about, including our little ways of doing things, and so much helpful advice to be shared. The family spirit and camaraderie can be time-consuming and I have often spent up to a couple of hours talking gun repairs; I know Lewis has too!

Lewis is a modest man who loves his work, but he is also a master in his profession, and his book is a must for all readers interested in the true art of gunsmithing. For the old gunsmith it will inspire a thousand memories; for the new gunsmith it will provide a good push in the right direction. I am sure that everyone who reads this book will be looking out for the next volume; I do hope we get one, as Lewis has so much more to tell us.

Chris Price, Master,
The Worshipful Company of Gunmakers, 2005

Preface

It gives great pleasure to be asked to write a book as, apart from a fortunate few, many talented would-be authors spend what must seem like a lifetime trying to convince agents or publishers of their literary worth. Yet at first I hesitated, for the gunsmith's trade is one of the last in which, to a certain extent, the principles of the craft societies of old hold sway. It is a world in which, at one time (like so many crafts), a man was sworn not to divulge his master's secrets; knowledge was power and could place you above other men.

This, of course, has its down side. Men of amazing skills whom I have known, and have even been fortunate enough to count as friends, have gone to their grave with more accumulated knowledge than many of us could hope to acquire in two lifetimes. They took their secrets with them and I know I would have been richer for a share of their wisdom. Even when a man does try and pass on his knowledge it is not possible either to give or absorb everything so, inevitably, with each passing generation we lose a little more.

I put my thoughts to some other members of the gun trade whose knowledge and skills I admired, and was pleasantly surprised to receive their unhesitating support for this project. The general view was that, in these times when gunsmiths often find themselves in a declining trade with, seemingly, the hand of Government against them, shared knowledge is a valuable support. Such a book might also help foster a broader interest, and this is important because the present insularity held so dear by some in the trade will surely only hasten our decline.

Added to these considerations was the thought that, given the rate of change away from practical skills, some information should be recorded before it is lost. Consequently, some historically significant techniques that are now little used have been dealt with so that not only the principles but the actual working methods have been detailed.

If this all sounds rather grand that is not the intention. What is recorded within these pages is my own working experiences – the work of a country gunsmith whose customer base varies from the Range Rover-driving Purdey owner to old Fred with his bucolic wit, bicycle and worn Stevens single.

L.A.S.P.

CHAPTER 1

Guns and Gunsmithing

Introduction

Gunsmithing as an occupation is probably unique. Its existence, whether described as 'Hainaulter', 'armourer', 'smith' or 'gunsmith', depending upon what part of history you examine, stretches right back in time to the earliest hand cannon of the fourteenth century. It has continued through periods of great technical change, from the matchlock to the wheel lock, the flint and percussion eras, and the overlapping use of all these types of firearms. The quantum leap in the nineteenth century was to the breech loader, which as a sporting shotgun was already perfected by the start of the twentieth century and has changed very little since. It is interesting to consider, when examining the latest creation from a maker of note, that when the first firearms were in use the longbow was at its peak and armoured knights rode to battle on horseback.

Today there are spy satellites and fearsome nuclear weapons, and some six centuries of progress have slipped by into history. Yet the sporting guns we use today still embody the same basic principles as those primitive hand cannon that first coughed their smoky challenge on some medieval battlefield. Fortunately, shooting as a sport attracts enthusiasts who strive to keep this history alive. Re-enactment groups firing matchlock and cannon provide a wonderful spectacle of family entertainment. Sportsmen and women still use flintlock and percussion muzzle loaders for competition. On a small English game shoot it is still possible to meet the black-powder breech-loading aficionado and other shooters of quite delightful, almost Edwardian eccentricity. Clay pigeon shoots vary from fun farmers' knockabout shoots, where socializing is more important than the score, to fiercely competitive prize events.

The gunsmith may be called upon to work on firearms from anywhere across this broad spectrum of our heritage. However, for the local gunsmith the bulk of work is on breech-loading side-by-sides and imported over-and-unders, the latter of which can receive some seriously hard work on the clay pigeon circuits.

The Gun Trade

Nowadays we tend to use the word 'gunsmith' to describe a repairer of guns as distinct from a gunmaker (a manufacturer), although in reality it is not that clearly defined. A gunsmith who has the requisite skills, or access to carefully selected subcontractors, may well build the occasional gun and as such, if he wishes, he may be entitled to advertise as a gunmaker. Many small gunmakers actually only do the part of the job at which they are particularly accomplished, and subcontract the other work or manufacture of parts. This makes good sense; it would be foolish for, say, a good stocker who also had other general gunsmithing skills suddenly to decide to make his own barrels without any experience. This, after all, was always the basis of the gun trade: the specialist skills available were used. Only the few larger companies might employ a variety of craftsmen to enable them to manufacture completely in-house, but at times of peak production even they would have to subcontract work to keep up with demand.

From the outside it would seem logical to assume that a gunmaker would carry out repair work or gunsmithing as well as new builds, and this is usually true. Most gunmakers have a workshop and carry out a variety of work, even if they choose to subcontract some of the new gun work to other specialists. As long as the work is to their standard of quality, specification and design, there is nothing wrong with this as it is the way of the manufacturing world at large; the integrity of the final product is their responsibility.

There are, and always have been, the individuals

who promote themselves as gunmakers but have no facilities, hands-on skills or real technical knowledge or involvement. In short, they subcontract the complete manufacturing process to others, sometimes abroad, and often their only contribution is the name. Similarly, a retail shop that has the word 'gunsmith' following its trading name may carry out some or all of the repair work or, again, subcontract it in its entirety. However torturous the route, the job usually finishes up in the hands of a gunsmith at the workshop bench. So, for the purposes of this book, we will accept the generally held view that the gunsmith is a repairer of guns capable of carrying out a broad range of repairs to lock, stock and barrel.

What does it take for someone to be a gunsmith? It is almost an article of faith in the traditional gun trade that an engineer, for instance, will never make a good gunsmith, and that a cabinetmaker will never be a stocker of any consequence (although I know of two who produce particularly fine work). There is some truth in such statements, but it is never possible to generalize and all too often such comments are simply a form of protectionism. So much depends upon the skills, knowledge and adaptability of the individual. While it is true that nothing illustrates the different approach better than the engineer reaching for his micrometer while the gunsmith uses the smoker for blacking down, really the two functions are complementary, and now quite formal engineering practices have come to the aid of the gun trade.

Much fuss is made about 'hand-built' guns; in fact, they are really hand-finished guns, the bulk of material being removed by machine. Cutting a large piece of metal away with a hammer and cold chisel is just hard work and has no bearing on the finished product. Many years ago the firm George Gibbs of Bristol used semi-skilled workers for this function, providing hardened filing jigs to work to. Advancing a stage further to the manually operated milling machine, it is then not much of a step, apart from the cost, to the computer numerically controlled milling machine – in other words a computer-operated machine. Parts can be produced that only require hand fitting and finishing, and this is where the skill lies. The finest example of this is the stock work, most of which is still done by hand to produce that look suggesting that the metal grew out of the wood. There is a lot of satisfaction lining out a stock blank and forming it into a delicate thing of beauty, but, realistically, cutting away large amounts of surplus wood by hand is a waste of time. No one suggests a stocker should really do without a bandsaw, neither was there criticism when Webley &

Scott used a stock copying machine for the basic shaping of their shotgun stocks.

From the gunsmith's point of view it means, with some makers, that accurate, standardized spare parts are available thanks to good engineering practices. However, what is admirable about the gunsmiths and gunmakers of old is that they produced exquisite pieces of work, often with the crudest of tools and almost entirely 'hand' technology. In spite of technological progress, many of these traditional skills are still required today. For most aspiring gunsmiths here lies the heart of the problem – how would the experienced gunsmith do the job? For me it was a long road. I repaired my first gun at the age of thirteen, a rather awful Chassepot bolt action twelve-bore. My next was a double twelve by Fletcher of Gloucester that required stock repairs to make it usable. In between time I amused myself by making miniature cannon which, with the carelessness of youth, I fired from my bedroom window sill into the chicken run down the garden. I never hit one of the fowl, who eventually got so used to this 'gunnery practice' that they took little notice. I remember, in my search for more performance, trying smokeless powder; I never did find all the parts of that particular cannon.

I hung around gun shops and gunsmiths to glean information and borrowed books from the library that have now become collector's items. I hoarded knowledge like a miser and eventually learnt, mainly by trial and error, to carry out simple repairs such as spring making, barrel dent raising, making strikers and recutting chequering. Years later I went to work with a gunmaker whose first advice was to tell me to forget what I thought I knew already because he would show me how to do the job properly.

I learnt that a gunsmith needs a variety of well-practised skills, an eye for the smallest detail as well as an appreciation of line and form and the ability to think in three dimensions. Then there was that indefinable feel for the job that is more artist than artisan. Most important was the knowledge, the know-how – the gunsmith's way. Although now very much a blend of ancient and modern, gunsmithing is still steeped in the practices and techniques of the past; real living history.

Variations on a Theme

In the simplest terms a shotgun is a smooth-bored firearm used for discharging multiple shot, which may also be used for firing a single projectile such as ball or rifled slug. The archetypal shotgun that most members of the general public will recognize is the double-barrelled gun with the barrels laid side by side – the most simple form of repeater. As a boy I rarely heard anyone refer to a 'side-by-side'; it was simply a double- as distinct from a single-barrel gun. It was only the rise in popularity of the gun with superimposed barrels, the over-and-under (or, as an old customer of mine insists, the 'up-and-under'), that led to the need for differentiation. Interestingly the over-and-under pre-dated the side-by-side, usually in the form of the turn-over gun, on which one lock served each barrel as it was rotated to the top position. The side-by-side was king in the nineteenth century and for most of the twentieth century, but now the over-and-under has overtaken it in popularity.

The single-barrel guns vary, from the simplest single-shot to more technically advanced designs such as the pump-action and semi-automatic magazine guns. The latter are sometimes erroneously described as automatic or auto, which they are not, as this would imply multiple shots from one squeeze of the trigger. If the semi-auto had been developed in the UK it would undoubtedly be described as a self-loading shotgun.

Combination guns that have both shot and rifled barrels are typified by the Austrian or German drillings. Most common are two shotgun barrels with a rifle barrel laid underneath. There are variations, including one shot, one rifle barrel in over-and-under format and, far more rare, over-and-under shot barrels with a rifle barrel laid alongside. Cape rifles, a type of firearm very much associated with the days of the British Empire, are side-by-sides with one shot barrel and the other (usually the left but not always) rifled. Another gun from this period is the paradox gun, a smooth bored gun with heavy, ratchet-like rifling in the last few inches of barrel. Of great rarity are multi-barrelled shotguns with three or even four barrels, which qualify as the ultimate collector's item.

There is a distinct limit to the number of ways barrels can be assembled, but not so with actions, which are many and varied. In that great era of invention, the nineteenth century, gunmakers strove to come up with something new, better or perhaps just different to get around someone else's patent.

Of the single-barrel guns, the most common are the break-open type with the barrels hinged to the action, typically the Harrington and Richards, Cooey and Astra Cyclops; despite their different countries of origin, these are nearly identical guns. A poor second are bolt-action guns; undoubtedly the most popular of these in the UK is the Webley and Scott .410, which is so often a youngster's first gun. The Marlin goose gun in twelve-bore three-inch magnum with box magazine is one of the longest-produced larger guns of this type. Both of these later designs are well advanced from the early Chassepot bolt-action black-powder guns, whose

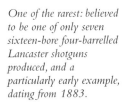

One of the rarest: believed to be one of only seven sixteen-bore four-barrelled Lancaster shotguns produced, and a particularly early example, dating from 1883.

11

Top: the unmistakable profile of a Martini action, in this case a twenty-bore converted rifle.
Bottom: one of the larger bolt actions, an American Mossberg.

least endearing features were an awkward cock-on-opening bolt, misfires and poor extraction.

A more elegant solution is the Snider action, which involves a hinged breech block swinging out and being drawn backwards for extraction. Once reasonably popular for large-bore wildfowling guns, this design originated as a simple breech-loading conversion for military muzzle loaders and has those desirable features of strength and simplicity.

Another single-barrel shotgun with military links is the Martini, originally Martini-Henry's – 'Henry' referring to the original form of rifling used. When converted to a shotgun, the Henry part of the original designation is not applicable so it becomes a Martini action. At one period it was quite common to convert ex-military Martini rifles to both .410 and twenty-bore; the latter, due to the constraints of the military action, was the largest cartridge to be chambered and that was at only two-and-a-half-inch (65mm) length. When Greener enthusiastically took up the design and made a commercial twelve-bore version, the well-regarded GP (general-purpose), later manufactured by Webley and Scott, was capable of taking a variety of cartridge lengths.

A fascinating array of single-barrel breech designs may be found on small Belgian 'Flobert' guns. Many of these, although smooth bore, were actually made to fire a single ball in a very short rimmed case. Although of very low power – the .22 version is particularly anaemic, and less useful than many air rifles – the 9mm shotgun version is quite

handy for ratting around the chicken run. These guns, in part due to the materials used but mainly because of the designs, are not suitable for conversion to centre fire. Also, due to the changes in British law in 1968, many fall below the minimum length of barrel to qualify as a shotgun in the UK, even though they are smooth bore.

The great majority of double-barrel guns, whether side-by-side or over-and-under, are of the hinged, break-open type. They are held shut with some form of locking device (of which more later), which, when released, allows the barrels to hinge open for loading or unloading. It is a type of action that has found universal appeal, used not only for shotguns but also the best double rifles. In the UK, breech-loading development as we know it, using a type of self-contained centre fire cartridge, dates from the 1850s. Prior to that there was the Lefaucheux pin fire and the Pauly of 1812. Rather typically, it seems, British gunmakers were tardy at first to develop these European advances. Once they got into their stride, however, they had, within forty years, perfected a style and design that became the world standard. There were blind alleys and strange variants, but a century that started with muzzle loaders finished with the perfected boxlock and sidelock breech-loading ejector gun that is still produced today.

It is, however, an immutable fact of life that someone will always try to find an improvement, so there are other double guns with different methods of breech closure. One of the best known, not so much for its availability but more for its curiosity, is the French Darne with fixed barrels and sliding breech. Even the idea of a kind of side-by-side Martini, which originally surfaced as a Money-Walker, has in more recent years been proposed again, although it has not been produced. Possibly the most prolific seeker of something different was P.V. Kavanagh, who seemed to revel in what might be described as the unusual alternative. As late as the 1970s this small company produced double-barrelled guns in which the barrels turned through ninety degrees and pushed forward to open, and a few on which the side of the breech hinged open, which became known as the 'book opener'. Its main production model had superimposed barrels that moved forwards on a tapered slide to open. At one stage Webley and Scott was in negotiation with Patrick Kavanagh with a view to adding it to its range – they did not, at the time, have an over-and-under – but negotiations failed and the gun continued to be marketed by Kavanagh as 'The Fenian'.

At one time a great many rifles were converted to smooth bore, either to qualify as the less legally restricted shotgun or because suitable rifle ammunition was no longer in production. The former relates mainly to twentieth-century military firearms such as the Lee Enfield and the latter to a lot of pretty little rook and rabbit rifles, spoilt by conversion to .410. Fortunately, as the interest in

vintage firearms has grown so has the availability, if not of loaded ammunition, then at least of cartridge cases, so it is now a popular move to reconvert them by sleeving with full-length rifled tubes. It was not just small rifles in break-open falling block, tipping block and any other types that were subject to this form of vandalism. Bigger rifles could be opened out for larger shot cartridges and elegant falling blocks such as the Field action adopted by Holland and Holland were targeted for conversion.

Locking the Gun Shut

The hinged break-open action is now, along with the top lever means of opening, the industry standard. The top lever is connected to a spindle (with some guns integral with the lever), which, when pivoted to one side, draws back a locking bolt in the manner of Scott's patent of 1865; the double-locking underbolt being a Purdey patent of 1863. It was not always thus and there are some interesting variants. The firm of William Powell & Son used, at one time, a well-regarded and now much sought-after lift-up lever with only a single locking engagement known as a single bite. The Westley Richards patent top lever, which is still produced, pivots on one side and draws back a top locking bolt that engages with a barrel extension. With early guns this was the sole

The 1970s rather different Kavanagh 'Fenian' action, the gun that nearly became a Webley & Scott. A striking example, fitted with Tungam barrels, a form of hydraulic tubing that is bronze in colour.

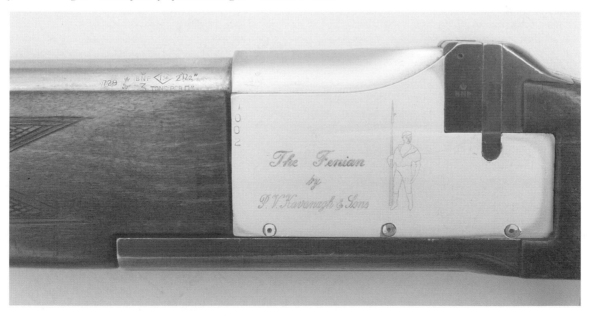

Powell lift-up top lever on a pretty little twenty-bore, a design dating from 1864.

These side-lever action Belgian-made .410 folding shotguns were once very common and a favourite with poachers.

and remarkably effective method of locking. Later guns are triple bite, incorporating also the Purdey-type double underbolt.

Push-down sidelevers are mainly found on economy .410s of Belgian origin with a simple locking method, but much more sophisticated versions may be found on double guns. Even in the cheaper forms it still produces unfailingly elegant lines, as well as a practical and handy method of opening. The underlever also had its day; one of the important historical milestones being the Murcott hammerless gun of 1871, which had an underlever wrapped around the trigger guard that also cocked the locks. Others include the Purdey underlever snap action at the front of the trigger guard, but the great long-lived design was the notably sturdy Jones underlever.

Henry Jones patented his important underlever design in 1859, and, although it was harder to make than some rival designs, it became deservedly popular due to its robust simplicity in operation and great strength. It survived on large-bore shotguns and big-game rifles for years after the top lever became popular on game guns. The Jones lever pivots out to one side and operates a double screw bite, which not only locks in place, but pulls the barrels down on to the action. It was used on a wide range of guns, from some of the cheapest to many of the most expensive models, the only criticism being that it was slow to operate, especially if mated to a gun with double-cocking hammers. Occasionally the Jones underlever may be found in a later modified form with a sprung underlever that snaps shut just as the action is closing.

The main component parts of Henry Jones' double-screw grip underlever, originally dating from 1859. This particular example is built into a Wilkinson patent Joseph Lang action.

Technically interesting Pountney double-bite latch-bolt action of 1878: easy to use but developed too late.

There were other means of holding a break-open gun closed that were practical and handy, if short-lived. The latch bolt fits crossways through the bar of the action and is pushed laterally, moving the bolt against a spring and releasing the barrels. Joseph Lang was one company that made use of this system but one of the best, although rather late on the scene, was the double-bite Pountney of 1878. Although it was doomed to obscurity, I know from practical experience with possibly the only example made that it was delightfully quick and easy to use. The great weakness is due to the large slot through the action. It would be ill suited to a hammerless gun and possibly somewhat weak for nitro proof.

Most over-and-unders stick to the well-tried principle of the top lever/spindle underbolt system, which, due to the constraints of the action, operate only a single locking bite, albeit across a

Valmet top-lock over-and-under, a type once imported to the UK by BSA guns.

very wide area. A notable exception is the Beretta, on which the top lever engages directly with the bolt, which locates halfway up the superimposed barrels. Less well known is the top lock system, where a kind of hood engages on lugs on either side of the top barrel; this was a system much favoured by Valmet and also used by others such as Miroku and Remington.

One thing is certain: if you dream up an idea for 'the perfect gun', it has probably been done before. The great thing is that from time to time those oddballs that occasionally surface keep the mind lively and in good working order. After a week or two of broken firing pins in cheap over-and-unders, worn-out unreliable ejectors and a run of barrel dent raising, I find myself ready for an interesting challenge, something the like of which I may never have seen before.

Relative Values

Ultimately the value of anything is what someone will pay for it, and sometimes logic does not enter the equation. The general consensus is that a sidelock is worth more than a boxlock, a trigger plate action lies somewhere in between and, in the very conservative world of shooting, oddball designs are for collectors. On the same theme, a famous-name gun should be worth more than a less well-known name, and expensive top-grade guns may be 'previously owned' while factory-produced items are simply 'second-hand'; it means exactly the same thing, but there is a social distinction. There are also some enthusiasts whose interest is confined to one maker and who may value its products above all others. Everything is relative to what someone will pay, and any attempt to link values can only be a guide, even for items all in similar sound and tidy condition.

What is rarely considered – because it does not necessarily have a direct monetary link – is the worth of a gun to an individual. A shotgun is, after all, in its most basic definition, a mechanical device for projecting lead shot at a (usually) moving target. If a gun, whatever its origins, fits an individual so well that it has become an unconscious extension of the body, effortless in use and gives a satisfactory high standard of success, it should be worth a lot whatever its market value. Alas, how many shooters have traded in their 'old faithful' for a better make or type of gun only to find that their standard of shooting, and therefore their enjoyment, plummets. It may feel good to be on a shoot with a fine gun bearing a well-known or perhaps even famous name, but if that gun was originally made for someone with quite different physical characteristics it will be more of a burden than a help. It will also bring a certain piquancy to the embarrassment of having the best gun and perhaps being the worst shot on the day. The lesson here is that there is nothing wrong with having a better-quality gun, indeed, there may be a lot of advantages, but it is more important to find a gun that fits, or to have it altered to fit properly, to gain any real benefit.

In monetary terms, it is best to forget that old imported single worn to a silver sheen as it is quite worthless (although it may be still capable in the right hands of dropping a pigeon or tumbling a rabbit dodging through the brambles). Even an immaculate Cooey single is almost unsaleable, and the aluminium-alloy-actioned BSA Snipe is something most shooters avoid. As for the single marketed as the Argyle, it was a brave try but surely a triumph of doubtful theory over good practice.

At one time most shooters in the UK had a single-barrelled gun tucked away either as a knockabout or hedgerow gun, perhaps a first gun kept for sentimental value. With the 1988 Act bringing in mandatory shotgun registration and the imposition of security devices, room in the gun cabinet was at a premium and many singles were scrapped or handed in during the Government 'amnesty' of the period. (Incidentally, most of the guns handed in were legally owned under the previous system of shotgun certification. Unsurprisingly, criminals proved themselves disinclined to hand in their illegally held weapons.)

The consequences of approved security requirements, even at a time when theft of shotguns was at an all-time low, very much depressed the bottom end of the market, and single-barrelled guns became virtually worthless. There are exceptions, including the Greener GP and the slim and stylish hammer guns of the late 1800s, but the only real value in a single is a big bore; this is an area where size really does matter – the bigger the better. There is something romantic about these big wildfowling guns, which recall mist-shrouded mornings and the call of wild geese. Sadly, the reality is that, with the ban on using lead shot over wetlands, most of these magnificent old fowling pieces no longer boom out across the salt marsh. In any case, in recent years it was never the real market – that belonged to collectors.

Regarding side-by-side doubles, the most affordable end of the market (ignoring hammer guns for the moment) was always the boxlock non-ejector but nowadays most game shooters prefer an ejector gun even if they do not need one for their particular sport. Best boxlocks were invariably ejector guns, although non-ejectors can be found built on 'best' actions, probably as special orders. Ejectors will often increase the value of a boxlock by as much as fifty per cent, and that is why it is still occasionally profitable to convert a good quality non-ejector gun to an ejector.

Side plates are fitted to add the style and elegance of a sidelock without the complication. They are no more than dummy lock plates and to some possible purchasers an affectation; there is also no doubt that, as some of the work is so neat, a number of guns have been purchased in the mistaken belief that they are real sidelocks. Done properly, side plates can transform the appearance of a deep, boxy action, especially an over-and-under, and give further scope for decoration. However, that is their only purpose and they are very much a matter of individual taste.

The same applies to chequering within a carved side panel behind the action of a boxlock. A plain panel with a carved drop point is probably the most desirable of this form of decoration. Next is the plain panel with oval form, and following on from that the panel that goes to a point, while the chequered panel (which serves no practical purpose) is,

Side plates on an over-and-under.

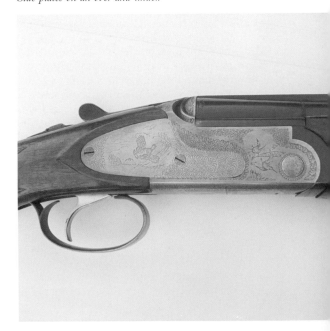

Fairly typical German-made drilling. A double-barrel sixteen-bore with 8 × 57 JRS (8mm rifle barrel) underneath, an example of a trigger-plate action.

to many potential owners, possibly one of the less desirable forms of decoration.

Sideclips – projections on either side of the breech that engage against the barrels – have been fashionable in the past and seemed more favoured on the continent, but tend to slightly depress the value of a gun. This partly because they are not necessary and spoil the lines, but mainly because they are a little unusual, and the market for good guns was always rather conservative.

Trigger-plate actioned guns are not common but are usually of good quality; one of the best known of this type is the Dickson Round Action gun, noted for its especially fine handling characteristics. In real terms this kind of gun runs second to the proper sidelock in the valuation stakes. As for those trigger-plate actioned drillings, because they have a rifle barrel they are, of course, a firearm and this always reduces the value in the UK because of the extra restrictions on ownership of rifled firearms.

A sidelock non-ejector is normally an economy, bottom-of-the-market gun with simple locks, but this is not always the case. In the past, not everyone wanted ejectors on a bespoke gun and early sidelocks often did not have ejectors. As a result, famous-name guns can very occasionally be found in sidelock non-ejector form. They do lag in value behind ejector guns because 'best' guns are expected by most shooters to have ejectors. Over the years a number of the later versions of these guns have been retro-fitted with ejector mechanisms, usually of the Southgate type. As for the basic, simple Baker-type sidelock, a plain non-ejector version

fetches about the same as, or a little more than, a comparable boxlock. Ejector versions fare much better and, with some reasonable engraving, fetch more than the average English boxlock ejector gun, but considerably less than a famous-name boxlock.

The sidelock ejector gun is the one to which most shooters aspire, so naturally it carries the most value, although there is a broad spectrum of prices within the type. Exquisite workmanship goes into these fine guns. As already mentioned, conservatism is everything and a bar-lock sidelock ejector gun is king of the shooting field. Single triggers and easy-opening devices attract extra value but the increased complication is not to everyone's liking. Back-action sidelocks, which many shooters regard as being visually less elegant, do not normally fetch the same money as a bar-lock gun. Unusual designs, such as the Lancaster Easy Opener, more often known as the 'wrist breaker', can sometimes be obtained for little more than the cost of a good boxlock ejector; this makes them something of a bargain, albeit they are an acquired taste.

As a young boy, the only over-and-under gun I had seen was a fine Edwinson Green sidelock made in 1912 (reputedly this design was the basis of Purdey's first over-and-under). Not only was this gun regarded as unusual and particularly beautiful, it would also now be towards the top end of price range, although still fetching much less than those produced by better-known names. It is also true to say that, even in this rarefied atmosphere, at the best end of the market over-and-unders fetch less than a side-by-side of the same quality.

The mass market for over-and-unders is dominated by the Browning B25, or what the Americans call the Browning Liege shotgun. The original John Moses Browning design, made in Belgium, combining the aspects of a practical, if slightly heavy gun with the flavour of old-world tradition, this has achieved something of a cult following. Indeed, it is not uncommon to find a lower-grade gun (this

The simple internals of what is commonly known as a Baker lock. Compare this with the 'true' sidelock (see Chapter 8).

really means a plainer gun, as they are all essentially the same workings) upgraded with better wood and engraving.

Miroku and Beretta vie very much for the same share of the market and the shooter's choice can come down simply to styling preferences. Dedicated competition guns, although initially some of the more expensive, are a little less saleable than all-round sporter-cum-game guns, and with a single trigger most users prefer a pistol grip to either a swan neck or a straight hand stock (which is rare with an over-and-under).

The interest in good-quality hammer guns has seen a revival in recent years, although many seem to be destined for export to the USA. Their chances of survival there are probably better than in the UK, considering recent events. As with single-barrelled guns, many hammer guns in the UK were handed in during 'amnesties' for scrapping off, some without even being valued. Some very fine guns, including antiques that did not need to be given up, were lost in this way. As a result, the survivors have seen a rise in interest, and, therefore, in value, but it is still very much a collector/shooter market.

Again, as with single-barrelled guns, the bore or gauge of a hammer gun has a considerable impact upon its value. The benchmark is the ubiquitous twelve-bore; the fairly rare short two-inch chambering is not the most popular because, although ammunition is available, it can be difficult to find a stockist. The twenty-bore has been very much in favour for the last twenty-five years or so and fetches a better price than a twelve-bore of similar quality. (However, for someone who cannot shoot with a twelve-bore, changing to a twenty-bore will be no help at all.) The sixteen-bore is the neglected Cinderella and yet it is a fine cartridge for game shooting, especially as many twelve-bore users now favour one ounce (28g) loads; the fifteen sixteenths of an ounce (26.5g) of the sixteen-bore is no real disadvantage. The advantages are a gun that is like a lightweight twelve-bore but with a bit more substance than a twenty-bore, an aid to most shooters, and at a cost that is lower than either of the others. Given time, interest in the sixteen-bore will increase and values should rise. For anyone wanting a good-quality gun at a reasonable price, now is probably a good time to buy a sixteen-bore.

The Americans have done much in rediscovering the twenty-eight-bore and promoting interest in this super little cartridge, possibly saving it from extinction. Traditionally, youngsters were introduced to shooting via the use of a .410, but a twenty-eight-bore would always have been a better choice had they been more readily available. Now, having risen from obscurity, a quality twenty-eight-bore can fetch serious money, even more on occasions than a twenty-bore.

The .410 is available in two-inch (50mm) fourten, two-and-a-half-inch (65mm) fourlong and three-inch (76mm) extra long or 'three-inch magnum' chambering. The bulk of .410s were cheap, starter guns; a great many for the two-inch cartridge, which is not as popular as the two-and-a-half-inch. Really good-quality guns are rare and expensive but it is something of an unpredictable market. Many .410s, perhaps because they were youngsters' guns, were neglected, so even at the more modest end of the market you have to be lucky to find a good one. Although a .410 is still one of the least expensive shotguns, unfortunately the cost of ammunition is high compared with twelve- and even twenty-bore. Consequently, many shooters regard this as being poor value for money.

Upwards from the twelve-bore is the ten-bore, again, like the sixteen-bore, an overlooked size without the cachet of its larger-bore brethren. The problem with the ten-bore is that, in most shooters' minds, it is not sufficiently bigger than the twelve-bore to make a worthwhile difference. It is true that the standard two-and-five-eighths-inch (67mm) chambered ten-bore was outclassed years ago by plenty of 'hot' two-and-three-quarter inch (70mm) twelve-bore cartridges, and compared with the three-inch (75mm) magnum, it is positively anaemic. Even the two-and-seven-eighths-inch (73mm) gun struggled. More recently, however, loads have been developed that bring it nearer the bottom loadings of the eight-bore. It is a pity it is not more popular, as a 'double-ten' is a gun of satisfying substance, an amiable light heavyweight that is easier to carry on a long walk down to the marsh than anything bigger. While not the most desirable in collector terms, when appearing in over-and-under form mated to a well-known name, that is entirely a different matter.

This leaves the eight- and four-bore guns. Of the two, the four-bore is the most desirable size but the eight is possibly the most used. The four-bore's attraction is that it represents the ultimate, although there were two bores that served as both shoulder and punt guns; fairly modern breech-loading punt guns have also been manufactured in two-bore. Of the shoulder guns, allowing for quality, they fetch more than any other size and new double-fours are once again occasionally being made.

Muzzle loaders represent a specialist field and a

great explosion of interest in the UK started about thirty-five years ago. Prior to that, muzzle-loading shotguns were a bargain, being much less desirable to collectors than pistols or rifles. For example, in the 1960s one of our local gunsmiths, who had a fondness for old guns, would restore muzzle loaders to shooting order. Singles started at £18/£20, doubles from £25. Now it is becoming increasingly difficult to find a decent gun and quite what has happened to them all is something of a mystery.

Fine flintlocks in good order are very hard to find and command some of the best prices, while drum and nipple conversions to percussion are looked down on as less interesting than a gun originally made with a percussion lock. Needless to say, drum and nipple conversions have been converted back to flint. Double guns, as always, fetch more than singles of comparable quality and, as with breech loaders, big singles are worth more than smaller bores and, as ever, well-known or recognized names count for a lot.

Customers often believe that the gunsmith has a personal collection of fine guns stashed away, but I am served well by a very good-quality breech-loading black-powder hammer double and a four-bore muzzle-loading single. If I were to add to those it would be a double 'flinter', preferably with inset locks. After all, some of us know in our heart of hearts that smokeless powder, hammerless ejector guns are really only a passing fancy!

The Gunsmith's Language

The first step on the road to understanding something of the world of the gunsmith is an appreciation of the names of gun parts used within the gun trade. At first, to the outsider it is a strange world where screws become pins, and occasionally nails; barrels have lumps; actions are made with fences; and hammers may have ears. The catch under the barrel that locates the forend is a loop, even if it never looks anything like one; on early guns at least the loop is a rectangular slot and the forend held by a crossbolt, which, of course, is nothing like a bolt that you might fit with a nut!

It is pleasing to think that, while they have probably changed subtly, these historic descriptive terms are still in use after many years. Nothing, though, seems to be carved in stone; there are certainly regional differences and, as this is something of an informal language, there may have been differences from one large gunmaker to another, or even between individuals. To confuse things still further,

modern factory-made guns use different descriptions, so, for example, while a boxlock over-and-under exploded drawing will show it to be equipped with 'firing pins', the same components on a traditional side-by-side are 'strikers'. Similarly, where a side-by-side ejector mechanism uses 'kickers', in an over-and-under these are more often 'hammers'. Still, at least this ancient language lives on in regular, if limited, use.

The Traditional Hinged, Break-Open Gun

Whether single, double, boxlock, sidelock or trigger-plate action, most of the same parts have the same or similar names. However, I have taken one small liberty in this book as an aid to clarity. Screws are usually referred to as pins but they vary in type, so I have described those having a full length of thread as screw pins, the type that have a parallel section and a threaded end as pins, and plain pins without head or thread as dowel pins. The location of these pins is also reflected in the description, as in hand pin or bridle pin. Those wood thread types (actually a tapered Whitworth thread) holding the tang to the stock are usually described as tang screws, a term that I have chosen to retain, although some in the trade would refer to these as pins.

Types of gun pin, below from left: pin, screw pin, tang screw; below: plain or dowel pin and roll pin as found on some modern guns.

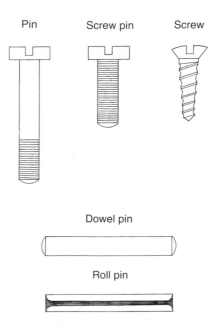

Pin   Screw pin   Screw

Dowel pin

Roll pin

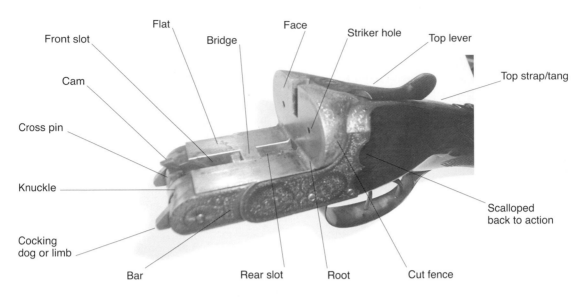

The main parts of an English Anson & Deeley boxlock action.

The Basic Action

The basic action is an L-shaped forging or casting consisting of the bar, which is the forward projection where the barrels fit, and the standing breech, against which the barrels close. At the rear of this is an extension in the form of the top strap. At the front of the bar is the knuckle, which fits into the mating part on the forend. In the middle of the knuckle is the cam (sometimes fitted in the forend), which pushes open the extractor, and behind that a cross pin fitted through the bar of the action as a hinge for the barrels. The two flat sections either side of the top of the action bar are, appropriately, the flats, against which the barrels sit. Between the flats, the front slot, bridge and rear slot are cut to accept the barrel lumps.

The area of the standing breech against which the barrels sit is the face, action face – at one time called the false breech – or breech face, drilled with striker holes and sometimes cut with vents (slots to allow gas escape in the event of a punctured or failed primer). The strikers are sometimes fitted behind discs screwed into the face of the breech, hence the description 'disc set strikers'. At the base of the breech face where it rises up from the bar is the root or radius; early actions were square-cornered and this part was always the root.

At the top of the standing breech there may be a piece cut to match the end of the top rib, which may be referred to as the shield. Either side of this lie the fences, which nowadays are usually ball fences (like half a sphere) or cutaway fences (in a 'D' shape). Guns followed a slow process of evolution and breech-loading fences developed from the raised fence around the nipples on muzzle loaders to help protect the user from the detonation of the percussion cap. Many early breech-loading hammer guns display beautiful raised fences that rival those on any muzzle loader. On some more expensive guns, gas vents are in the form of hollow pins – vent pins – which may serve the additional purpose of locking in place the striker discs.

The top strap at the rear of the standing breech is sometimes referred to as the tang, but it is the top strap when there is not a safety incorporated, and only the reduced part of the top strap when fitted with a safety is a tang. The safety may be described as a safety button, a safety catch or simply a safety. There are also side-mounted safeties, usually of the Greener design.

In terms of the internal mechanism, a boxlock contains hammers that may have fixed noses or separate strikers. The cut-out where the sear engages is the bent, and that part of the sear that contacts the bent is the sear nose. The sear, like the hammer, has a pivot through which is fitted dowel pins, also variously described as action dowels, sear and hammer dowels, or plain action pins. The sear extends back into a leg with a peg set at right-angles, which is the point of contact for the trigger blade to lift when the finger piece or trigger is pulled.

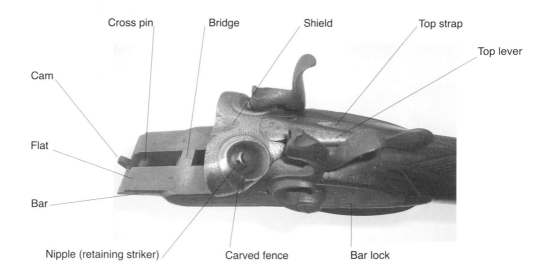

Cross pin · Bridge · Shield · Top strap · Top lever · Cam · Flat · Bar · Nipple (retaining striker) · Carved fence · Bar lock

Hammer gun with deep carved fences and nipples retaining strikers.

To cock the mechanism and compress the mainspring there are cocking dogs, sometimes called limbs, which project through the knuckle to engage into the mating knuckle of the forend iron. If it is an ejector gun, the cocking dog may be cut to incorporate a trip to operate the ejector sear. Alternatively, ejector trips in the form of rods or plates lying alongside the hammer may be fitted.

Boxlocks invariably have a bottom plate held on by a screw pin to cover the lockwork lying in the action slots. Back from that is the trigger plate, held in place with a front screw pin and, most often at the rear, a hand pin. Part of the trigger plate is the box, which connects with the action or breech pin and is the pivot point for the trigger blades.

The two common forms of safety mechanism are the push-up safe, used for economy purposes, and the rocker or dickie-bird safe, used on more expensive guns. With the latter the rod that connects the top lever with the safety button is the safety slide or range. This connects to the crank with a simple form of articulated joint that links it to the rocker, which in the 'on' position locks against the trigger blades.

With the simpler, push-up safe the range is a vertical bar that fits into the base of the safety button, formed at right-angles with a rod that operates the safety from the rear of the locking bolt.

The locking bolt, or simply bolt, slides in the bar of the action below the flats and between the action slots where the hammers, and so on, sit. It is

connected to the spindle by an offset lug, the spindle running through to the top of the action behind and between the fences and connected by a square to the top lever. The top lever in turn is secured to the spindle with a top lever screw pin.

The vee mainspring is nothing complicated – the sears are spring-loaded by means of flat sear springs, the top lever by a top lever spring and the safety with a safety spring.

The Forend Iron
The forend iron is comprised of the knuckle and the end of the top face is the toe or nib. The long bar to which the forend is fixed is the stale or steel and, if an Anson & Deeley pushrod mechanism is fitted, the front of the stale is the tube or pipe. With the Anson & Deeley system, the forend knob that is depressed to remove a forend is part of the rod that locates on the bolt that locks the forend in place.

The Barrels
The barrels have two lumps (occasionally called steels) underneath, between the barrel flats. The front lump has the half-round cut-out of the hook and at the back there is a bite where the bolt engages, and a semi-circular profile known as the run up. The similar but concave face of the rear lump is the draw, or circle, and at the back of the lump is another bite. The ends of the barrels that sit against the face of the standing breech are

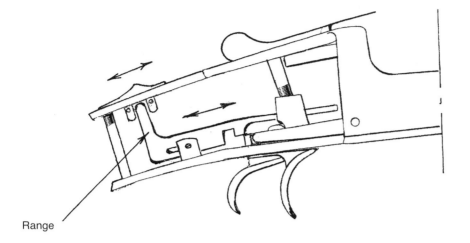

Range

One of the two most common safety mechanisms on side-by-side shotguns: the economy push-up safety.

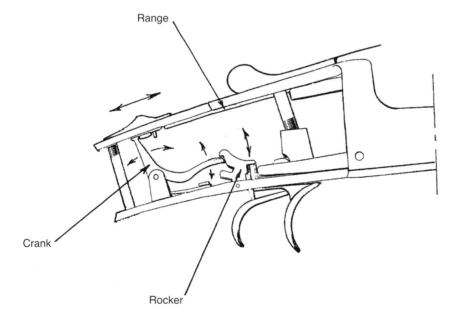

Range

Crank

Rocker

The more expensive 'dickie bird' system.

known as the barrel faces or breech ends and between is the extractor, which may be one piece, as on a non-ejector gun, or split in two, as on an ejector gun. The cut-out that accommodates the extractor is the extractor bed and the leg of the extractor fits into the extractor hole. There are several ways of keeping the extractor in line: it may be conncted into a cut-out in a barrel extension or it may use a peg that fits into a second extractor hole. This may be a single peg (non-ejector) or separate pegs (ejector gun) or, in the case of the latter, a split peg formed as part of the extractor.

Barrel extensions that engage into the top of the standing breech are most commonly doll's head (which appear shaped as a head and neck), Greener cross bolt, or as a third bite engaging under the top lever and sometimes hidden until the gun is opened.

The top rib may be concave, flat or Churchill type (raised and tapered). At the front of the rib is the foresight bead and at the rear where the barrels are held together it may be a laid-over rib, which is the full length of the barrels and a feature of best guns. Better ribs also have vee-shaped ends that fit between the barrels at the muzzle, sometimes

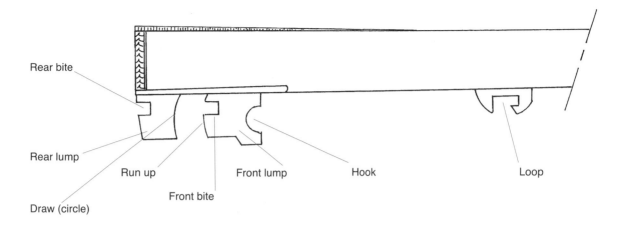

Rear bite

Rear lump

Run up

Front bite

Draw (circle)

Front lump

Hook

Loop

The relevant parts of a hinged break-open barrel.

referred to as nose ends. Cheaper ribs are finished by plugging the gap with solder. At the business end there are, of course, the muzzles, then underneath, the bottom rib, forend loop and keel rib, sometimes called the quarter rib. In the better designs, the keel rib is joined to the forend loop and is then almost always referred to as the keel plate.

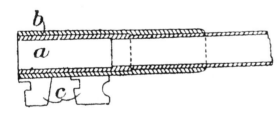

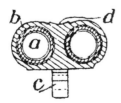

Monks 'Monoblock' 1881

W.H. Monks patent of 1881, which states how the breech and lumps are made in one piece and the barrels sleeved in from the rear of the breech piece. Monks recommended brazing or soldering in place or, as an alternative, shrinking the breech piece on to the barrels.

Barrels are made from either steel or Damascus steel (a mixture of iron and steel forge welded in the form of a helical strip). Steel barrels are finished blacked while Damascus steel is browned to bring out the contrasting patterns produced by the alternating bands of darker iron against the lighter-coloured steel. Sometimes the broad use of the word Damascus is improved upon to describe the individual type of pattern produced by this manufacturing process. Therefore we have single iron, two iron and three iron stub Damascus, as well as a variety of other names. The simpler and cheaper way of manufacture is scelp, which appears as a simple twisted pattern sometimes called Damascus twist. Single iron Damascus can appear as a rather broken pattern, which gives rise to the description 'puddled', which I do not believe to be correct.

With a double gun, barrels may be described as chopper lump. This indicates that the lumps are made in two halves as extensions of the barrel tubes and look rather like choppers, or small axes. Another method of double barrel assembly is to dovetail the lumps in between the barrel tubes; in both cases the breech assembly is brazed together. The most economic method of assembly is to spigot the tubes into a machined block that forms the breech part of the barrels and includes the lumps. This is, in essence, the same as barrel sleeving as a method of repair. Just to prove that little is new, a similar method, albeit inserting barrel tubes from the breech end, was patented by W.J. Monks in 1881, the same system, I believe, as used on the French-made 'Robust' gun.

23

Stock and Forend Wood

The forend may be either a splinter (narrow) forend or, less commonly, a wide beavertail type, which wraps partly around the lower side of the barrels. The bottom of the forend that rests in the hand is the belly. The front is the tip, which may be plain wood or fitted with a forend tip formed from horn or steel.

The stock has a head where it fits against the action. To the sides at the head of the stock there may be carved panels, some with a small carved end known as a drop point, tear-drop point or, to some older hands, a bottle point. Further back is the hand, where the stock is held, and the grip, that is, the chequering. Occasionally, the hand is described as the wrist! Back from the hand is the raised comb, along the side of the comb is the face, and at the end of the comb is the bump or heel. The shoulder end of the stock is the butt, the position at the middle of the butt is the middle, and the bottom end is the toe. The butt may be made of plain wood or fitted with a heel plate, butt plate or butt pad. The term heel plate has been used to describe both a butt plate made of steel or horn (or other material) or a butt pad. Additionally, there are delightful small fancy plates fitted to the top and bottom of the butt, variously described as tip and toe plates, quarter plates or toe and heel plates.

The underside of the stock may be straight hand – as in a straight line – or in a shallow curve to form a swan neck. Alternatively, it may have a short, curved grip that is rounded on the end (bag grip) or a grip with a flat grip cap. That final component, the trigger guard, is formed from the bow that wraps around the triggers, which may have a rounded edge, known as a rolled edge, the bow extending into the tang or tail, which lies along the stock. Sometimes there is a delightful piece of silver or gold let into the stock for the owner's initials; this is an escutcheon if shield-shaped or an oval if oval-shaped (actually nearer to round until it is fitted with the curve of the stock).

The Sidelock

The basic essentials of the sidelock differ little from those of the boxlock. The levers that cock the locks are lifters rather than cocking dogs, and the lockwork is mounted on lock plates rather than inside action slots, which are not present on a sidelock; neither is there a bottom plate. This lockwork consists of a hammer that is formed as part of the tumbler that engages on the sear, both of which are held in place by a bridle, which, if cut away, is a pierced or ring type. The tumbler is connected to the main-

spring by a swivel, sometimes described as stirrup, and the whole assembly is fixed to the lock plate with bridle pins. The locks are secured in place with either a single side nail or lock pin or including further screw (lock) pins behind the fences.

Better locks are fitted with an intercepting sear, which may simply be referred to as the interceptor and, of course, both main sear and intercepting sear are operated by small curved or vee springs.

Sidelocks are either bar locks or back-action locks. With the bar lock the projection on the front of the lock plate is set into the action bar with the mainspring located behind it, while the side panel of the stock is cut away to accept the rest of the lockwork. With a back-action lock the lock mechanism and the mainspring are contained within the cut-out in the stock. It is not always obvious that a gun is a back-action lock as some are the shape of a bar lock. These are of varying complexity but the usual identifying factor is that they do not show the end of a mainspring peg through the extension on the lock plate that lies in the bar and are back action bar locks. Others are obviously back-action locks as the lock plate engages only a short way into the action, or is completely behind it.

Some argue that back-action locks weaken the stock due to the amount of wood cut out. If that is the case, then so do bar locks, as the only difference is the cut-out for the mainspring. Back-action locks that engage into the bar may be marginally stronger as they form, in effect, a support plate from the action. The advantage of this type of back-action lock is that there is no cut-out in the bar of the action for the main spring. This is why many powerful doubles have back action bar locks allied to a long top strap, which, in the case of double rifles, may be carried up and on to the comb.

Hammer Gun

Hammer guns are mainly either bar or back-action sidelocks, although there are cheaper hammer guns that qualify as a form of boxlock with the hammers emerging along the centre of the top strap. Those that are sidelocks are little different from the hammerless – or internal hammer – derivations, except in that the hammer is separate to the tumbler. The pivot of the tumbler projects out of the lock in the form of a square, which engages in the square in the hammer, which in turn is held in place with a hammer (screw) pin. The curved front of the hammer is the breast or belly (distinctly a beer belly on some), which runs into the curved neck. The front of the hammer that hits the striker is the nose and above that is the ear or thumb piece. On hammer guns where the sprung striker is held in place with what looks like a kind of nut, this is still referred to as the nipple, the same as the item on a muzzle loader over which the percussion cap is placed. Remember, guns evolve quite slowly, as do the names of parts. In an age of unreasoning hurry, the world of the gunsmith can be an oasis of sanity.

Hammer gun locks are either double cocking or rebounding locks. In the former, the safe or half-cock position is engaged manually with a deep cut bent in the tumbler, which is designed not to let the sear disengage if the trigger is pulled. It is just the same as a lock on a percussion muzzle loader and the identical principle to the half-cock on a flint-lock, which is where it got its name. In a rebounding lock, the end of the mainspring is extended and, after the hammer's full travel or hammer throw, this part of the mainspring pushes the tumbler back just past the half-cock position. This sprung half-cock is not as far back as the manual type even though the main spring holds the tumbler back past the half-cock position.

The type of back-action lock on a hammer gun that lies a long way down into the hand of the stock (not to be confused with a hammerless lock) may be a potential weakness, due to the amount of wood on opposing sides of the stock removed to accept the lockwork and quite long mainsprings. Interestingly, when accidentally subject to sideways forces, such a stock will usually break at the back of the locks, more often than between the locks where the stock is cut away to clear the lockwork.

Modern Terminology

Modern factory-made guns not of British origin use their own terminology. Listed below are some of the old British names and their more modern counterparts.

Breech or action pin	Frame screw
Top lever	Top snap
	(also used in the 1880s)
Hand pin or long hand pin	Rear frame screw
Striker	Firing pin
Ejector kicker	Ejector hammer
Action	Frame
Range (push-up safe)	Safety square
Cam	Extractor lever
Pin or cross pin	Axis
Bottom plate	Frame plate

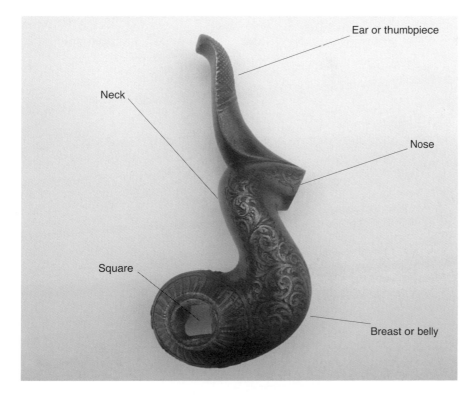

A very elegant hammer off a William Fletcher of Gloucester, an early underlever snap-action twelve-bore (my own second gun).

Ear or thumbpiece

Neck

Nose

Square

Breast or belly

There are also some differences in terminology between the Birmingham and London trade:

Hammer	Tumbler
Chamber	Breech
Bent	Hammer notch
Cocking dog	Lifter
Foresight	Bead
Drop points	Tear drops
Extractor	Ejector
Intercepting sear	Safety sear
Forend	Forestock
Mainspring	Hammer spring

This gets a little more confusing when firms such as Browning use many of the traditional part names mixed up with a few of their own. Therefore, while the top lever is still a top lever, not a top snap, the spindle is described as a top lever dog shaft. Some over-and-unders have hinge discs through either side of the frame, which act as a short, two-piece cross pin. Split extractors are ejectors and the forend iron may become the forend bracket. Where

the spindle is formed as part of the top lever it is referred to simply as the top lever. The action is more often described as the frame or, occasionally, the receiver, which is the common description used with semi-automatics and pump actions.

Descriptions of pump-actions and semi-autos are a long way from the traditional terms of Birmingham's Steelhouse Lane, once very much the heart of gunmaking. Breech bolts or breech blocks, carrier assemblies, cartridge cut-offs, forend liner assemblies (the same as action slide assemblies) – these parts will be dealt with in more detail in Chapter 4.

The gunsmith's language is both ancient and modern, then, varied and sometimes confusing, but with some terms that go across the board: for example, to most gunsmiths in the English trade an over-and-under firing pin is still, and always will be, referred to as a striker. Covering such a wide and varied vocabulary would fill a book on its own, but such a book, although probably a useful reference, would perhaps not be a best-seller. (See Glossary, page 163.)

CHAPTER 2

Tooling and Equipment

Introduction

Gun repairs vary from the simple replacement of factory-supplied parts, which can be done on the kitchen table, to the more complex jobs, such as relaying ribs or major stock alterations. Whatever the job, it will almost always require some sort of tools.

There are some jobs that are relatively simple. Many gunsmiths will say that they have lost count of the number of times ejector shotguns have been brought in to them because 'the forend will not fit on'. Inevitably the ejector kickers are found to be in the uncocked (fired) position. This makes it very difficult to snap the forend in place while providing enough force at the same time to cock the ejector kickers. It also puts a strain on the forend. The trick is to either cock the ejector kickers separately with a piece of brass bar inserted into a file handle, or, with the barrels detached, push out the extractors, fit on the forend and push the extractors against the bench to cock the ejectors. Not everything is this easy, though. If it were, everyone could be a gunsmith.

This chapter covers a variety of tooling, from proprietary hand tools to suitable machinery and tools peculiar to the art of the gunsmith. Gunsmith-made tools that have regular use or a variety

A typical gloomy old workshop for a gunsmith – it was lighter before the undergrowth grew over the window.

of uses are dealt with in detail here; specialist tools for one type of job are listed, but detailed elsewhere, in the relevant chapter dealing with their particular use.

The Workplace

The workplace area relates directly to the type and amount of work to be undertaken. For the single-handed gunsmith intent on little more than servicing and minor repairs, a floor area of no more than twelve square yards or metres – about the same as the average room in a modest house – can be quite sufficient. It is, though, worth remembering that you always seem to need more room than you first estimate, especially taking storage into account. As the type, scope and complexity of work increases so does the need for a larger workplace. In addition, there are some functions, such as blacking and browning, which, due to the processes used, should be kept separate from other work. No matter how modest it is, machinery is also best separated by at least a little distance from any cleaner handwork area.

Ideally any workshop, regardless of the function, should be warm in winter and cool in summer, with plenty of natural daylight. This is starkly in contrast to the dingy rooms used by many trade workers at the height of the English gun's popularity.

For both the full-time and part-time gunsmith, it can be a problem deciding what to do with customers when they visit. In a world obsessed with health and safety regulations, great care must be taken not to let them injure themselves out of curiosity. Discretion is also an issue: you do not want someone peering at a job on the bench to see that a shooting rival is having modifications carried out to his favourite gun. Customer confidentiality is of considerable importance.

When a workshop is housed in an annexe to a retail business, there is often a counter or hatch, through which the customer's gun disappears into a secret world, eventually re-emerging in full functional order. Even some businesses that are no more than a husband-and-wife concern manage a small separate reception area that also serves as an office. A few sporting prints on the wall and, perhaps, a display of old shooting accessories will create an environment in which the customer feels comfortable.

In a simple one-room workshop, one solution is to have a small gated area or hinged counter to separate the customer from the workplace proper. The old gunsmiths had a simple method – always sup-posing customers were even allowed in the work-shop – which involved a large piece of grubby rag on the bench. If the customer was given permission to enter, so the rag would be drawn across the job in hand, hiding it from prying eyes.

Whatever method is used to separate customer and work, there is one certainty: the gun owner is an enthusiast, who will find the gunsmith's work-shop a place of endless fascination. Let him in, especially on a day of inclement weather when there is a warm stove in the corner, and it may be quite a job to get him to leave before closing time.

Basic Tools and Equipment

The tools a gunsmith needs will vary according to the type of work to be undertaken. His toolkit may contain just a selection of screwdrivers, punches, files and a handful of woodworking tools – a keen amateur model-maker will already have much of the necessary equipment. At the other end of the spectrum is a fully equipped workshop complete with lathe, milling machine and associated equipment.

While it is not within the scope of this book to detail every tool that might be used, or how to use them, the following covers a selection of essentials.

The Workbench

The workbench is the least-considered but most-neglected and abused piece of equipment in any workshop, yet it is used every working day. Comfort is important, and it soon becomes obvious that one size does not fit all. If anything, it is better to have a bench slightly too high than too low. With a tall bench a duckboard can always be used to obtain a comfortable working height, but a bench that is too low cannot be adjusted in this way. There is nothing worse than being hunched awkwardly over your work and, in the long term, it may invite back problems. Ideally, when standing upright, you should have the top of the vice about level with the point of your elbow.

A workbench should be as solid and sturdy as possible. Mine is built against a wall and under the window for best light conditions, so I do not work in my own shadow. The main frames are fixed at either end on brick supports. Intermediate legs are no further than 48in (1.2m) apart and all main timbers at least 4 × 4in (100 × 100mm) section. There is no point in having a heavy frame with a flimsy top, so 2in (50mm) timbers can be rebated into the top frame and screwed into place. Oil-resistant

A sturdy workbench with plenty of natural daylight. Note the sprung soft jaws on the vice.

hardboard or similar sheeting can be used to produce a good working surface. The top of the bench also slopes just a degree or two towards the user, so that spilt fluids can be easily mopped up rather than running down the back of the bench.

The Vice

The leg vice (often called a blacksmith's vice) is much associated with gunsmith use and still has a role to play, but there are better options. For stock work, particularly when dealing with a blank that needs to drop well down into the vice, the leg vice has few equals. For precision work and most general functions, the engineer's vice is far better. It is most useful when fitted with a quick-release mechanism, and if the jaws are not reversible, they should be ground flat and true. There are times when it becomes necessary to hold a job without soft jaws. Flat, precise jaws will, with care, not mark the work piece; the serrated evidence of the vice's standard grip is the unmistakable sign of the rank amateur.

For maximum rigidity the vice should be mounted adjacent to a main support leg of the bench, with the front edge of the rear jaw in line with the edge of the bench, to minimize overhang. When holding a pair of side-by-side barrels by the lumps, this arrangement necessitates a cut-out in the front edge of the bench.

Vice Soft Jaws

Soft jaws (sometimes a relative term) are usually essential to place over the steel jaws to avoid marking or damaging a job, as often the gunsmith will be working on a finished item. Thin brass or copper is useful when the need arises to hold a small item very rigidly. For general-purpose use lead is unbeatable and moulds are available to cast your own. If you do not want to go to this amount of bother, roofing lead can be folded to fit the vice jaws; while not quite so good, this is a practical substitute.

Wooden protective jaws see a lot of general use. They are best when made with legs that act as springs so, as the vice is opened, they open with it.

Mine are faced with thin cork and, when a finished stock is to be held, I also insert a thick piece of soft leather for added protection.

Bench Horse

The bench horse is a raised support fitted to the bench adjacent to the vice. It is most pleasing when made of hardwood to a pattern that has evolved over many years, but it may also be made of wood or steel. The important requirements are that it should be about 2in (50mm) below the top of the vice jaws, project about 4in (100mm) forward of the edge of the bench, and should be able to swivel, to accommodate a job at an angle, or to be moved out of the way. The upper surface, which makes contact with the work, should be covered in a reasonably soft material such as felt or leather.

The distance the bench horse is set from the vice is a matter of practical choice, usually between 12 and 18in (300 to 450mm). Most gunsmiths will use a bench horse on either side of the vice, but it is not unknown for only one to be used.

Turnscrews and Screwdrivers

A turnscrew is what the old-time gunsmith and the present-day traditionalist call a screwdriver. Turnscrews, as a distinct type of screwdriver, are still available, often as beautifully made cased sets produced by accessory manufacturers. As collector's items and for occasional use they are fine and certainly interesting. For general workshop use, the

The blade of a gunsmith's screwdriver or turnscrew should be shaped in a particular way.

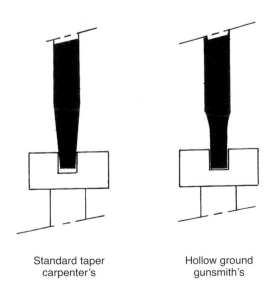

Standard taper   Hollow ground
carpenter's   gunsmith's

gunsmith needs something far more practical: a screwdriver that has the shank running right through the handle, which is either heavily serrated, square or hexagon in form to get a good grip, and of varying lengths. Sometimes you literally need to 'get your shoulder' behind the job.

Everyone seems to end up with a motley selection of screwdrivers acquired over the years. The one important factor common to them all is the shape of the blade. Unlike a carpenter's screwdriver, which has a tapered blade, the gunsmith's should be parallel, to engage the screw slot forming an exact fit. Sometimes it even becomes necessary to modify an existing screwdriver to engage properly into an important screw. To a carpenter, this may seem a long way around the job, but for the gunsmith that is the way it is – there are no short-cuts. Gun screws are often fitted very tightly so, if the screwdriver blade is ill-fitting, and bears on each end of the blade rather than across the full width, it is likely to break.

Punches

There are plenty of proprietary punches on the market. Most have flat ends, but occasionally they can be found with rounded ends, or with concave ends, which is particularly useful. Inevitably the gunsmith finishes up making punches out of scraps of steel or brass bar. When a punch is slim (which is the norm) it is advisable to have a heavy handle. This adds considerably to rigidity and avoids a nasty stinging vibration when striking with the hammer. In the worst instances, a reluctant pin can literally numb the fingers.

A starter punch has a long tapered nose for ultimate strength and rigidity. It is invaluable for obtaining the initial movement on a sticky pin and can save bending the longer thin, parallel punch.

For most general maintenance work there is one golden rule: if you have to hit the punch really hard, there is something wrong! Brute force should never be necessary to strip a gun. If a part does not want to move, always check and look for the locking screw or cross pin that may be holding it in place. It should not be necessary to use a hammer or mallet bigger than 4oz (100g) to drive a punch.

Hand Tools: Metalworking

File To the uninitiated it can come as quite a surprise to find that, not only does the humble file have different grades of teeth, but it also comes in a plethora of shapes and sizes. It is not difficult to recognize a square- or round-section file for what it is. However, a triangular-section file with equal

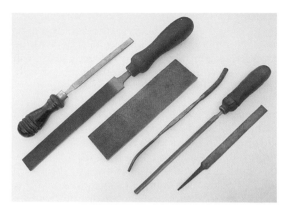

A selection of specialist files, from top, left to right: diamond file, chequering file, draw file, riffler file, three-square dovetail file (teeth only on one side), screw-slot file (teeth only on the side).

sides is a three-square, an obviously flat file a pottance or hand, and a file described as flat has curved sides. Then, to make it even more interesting, there are pillars, equalling, warding, knife, barrettes and sinkers, rat-tails and delightfully named rifflers – a generic term for a whole family of strangely shaped files.

There are more, including the real specialists such as draw files, chequering and screw-slot, and Abra files, which are fitted into a hacksaw for use. Sizes can range from as large as 14in (360mm), for heavy hand-filling tools, to the slimmest, most delicate needle file. The teeth, which are correctly called 'the cut', range from rasping dreadnoughts to tiny teeth (described as smooth), which are difficult even to feel. For extra hard surfaces there are diamond files, which are actually made from diamond dust set in nickel. Of course, not everyone needs all types of file, but it is useful to know what is available.

Most files are not supplied with handles and only a masochist who wishes to impale their hand on a file tang uses one without a properly fitted handle. Synthetic file handles are available, but probably the most common are still the wooden type in sizes ranging from 3in (75mm) to 6in (150mm). New ones can be fitted quite easily by holding the file body in the vice, heating the end of the tang to a glowing cherry red and, with a gloved hand, pushing the handle into place. This operation is accompanied by a certain amount of smoke and flame and, as that dies away, a couple of sharp taps with a lead or wooden mallet will produce the final fit. Attempting simply to drive a

wooden handle on to a tang often results in the wood splitting.

Needle files are usually formed with a parallel handle of small diameter. They are more manageable when fitted with a collet-type handle that provides a larger, more comfortable grip.

It is easy to take a file for granted and view it as a fairly basic tool to be shoved under the bench when not in use. Such ill treatment will, however, shorten its life. Whether you are operating as a part-time or full-time gunsmith, tools have to be purchased and you can only pay for these items out of the profits. Each time a tool goes down it creates a hole in your income. As a young man I learned a lot from the old chaps (as I called them when I did not know better), who took the greatest care of all the tools with which they made their living. Even the humble file benefited from their attention to detail. It was second nature when picking up a file to tap it against the edge of the bench to make sure the handle was tight before starting work. The file was never dragged back across the job, but lifted slightly to clear the work on the backstroke and, every so often, it was knocked on its side against the wooden bench to clear loose filings from the teeth. If the file was new or the material subject to pinning (the build-up of metal in the file teeth), chalk would be rubbed on to it. Even cleaning was quite a little ritual. The favoured method was a piece of flattened copper pipe run across the teeth to clean out any debris – more effective than any wire brush.

Hand Tools – Woodworking

Chisels and Gouges The straight chisel is the staple woodworking tool that can be used or adapted for a number of functions, from paring off wood to carving, scraping and cutting. I prefer the flat, or English, chisels to the later bevel-edged type and modify the angle of the cutting edge to around twenty degrees, rather than the general standard twenty-five to thirty degrees. Most of the time, on gunstock work, a chisel is used for what is effectively woodcarving. The shallow-angle blade is more suited to this type of work, although a cabinetmaker will probably tell you that for hardwoods, such as walnut, a steeper-angle blade is better and a fine angle is normally used for soft woods. The best thing is to find out what suits you, based on actual working experience.

It is not really possible to produce this fine-angle blade with a hand stone unless you have the patience of a saint and a lot of time to waste. If you do not have a grinding wheel (and this amount of reshaping should be done on a 'wet' wheel), then it makes sense to subcontract the work to a specialist tool-grinding company. It is also worth noting that, after grinding, the chisel is honed and lapped to give the final cutting edge, which is best made a few degrees steeper than the main angle; this makes for a stronger cutting edge. One useful tip to ensure a chisel is sharp is to hold it up to the light. A sharp edge does not reflect light.

Once you move away from the simple world of the straight chisel there are other variations, such as straight corner, skew, side chisels, foot chisels and doglegs – quite a range, but perhaps less confusing than the huge variety of files available. However, some of terms used to describe the gouges, many of which probably go back centuries, seem designed solely to confuse anyone perusing a catalogue. Gouges may be bent or curved, back bent or long pod; more strangely, they may be alongee, veiner or spoon. The confusion only grows worse in the pure carving tools where terms such as shallow lift, extended short bent, standard forebent, upswept and bent skew are encountered.

Fortunately, for stock repairs (rather than stocking) only a few chisels and gouges are needed, and with a little experience it soon becomes obvious which are the most useful; however, if any of these types of tool become available at the right price, it is worth buying them. Apart from those that I inherited from my grandfather, most of my wood-

working tools have come from old-fashioned junk shops or car-boot sales, or have been scrounged from friends. It is surprising what good-quality tools sometimes lurk under a layer of rust and grime and, with sympathetic restoration, still have best part of a lifetime of careful work in them.

Spokeshaves Next to the chisel, one of my most frequently used woodworking tools is the spokeshave. Gunstocks exhibit subtle changes of shape and curvature that are particularly suited to this tool. I still use the wooden-bodied type, which offer a feel that seems to be lacking in those made of cast steel. Not only that, but the old wooden spokeshaves also come in a variety of sizes, from large tools that will move wood almost like a plane, to those approaching pencil dimensions, which are ideal for delicate use and around areas such as inside the curve of a pistol grip stock.

Modern Woodworking Tools It is not unknown for an element of snobbishness to creep into the world of quality woodworking. The practical woodworker uses whatever is appropriate to do the job in hand, and usually a tool with which he feels comfortable. However, some workers will decry the use of tools such as rasps, suggesting their use is inappropriate on a good gun. Some gunsmiths even make sure that the rasp reappears from under the bench only after a customer leaves! Realistically it does not

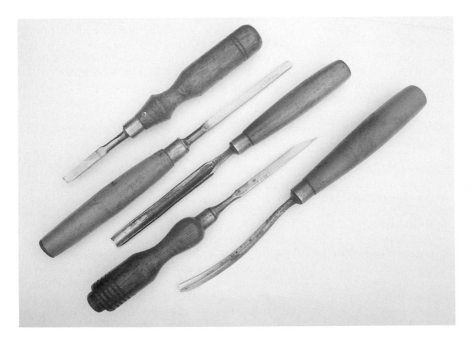

A few of my favourite woodworking tools, from top, left to right: straight chisel worn short from much resharpening; chisel modified to a round-nose scraper, useful when fitting forend wood to the underside of barrels; plain part-round gouge; straight chisel sharpened on the side for narrow cuts; curved vee spoon gouge.

matter what tool is used as long as it will do the job – it is the quality of the finished item that is important.

Rasps are available in a variety of shapes and sizes, relatively cheap to buy and usually with ready-fitted handles. For jobs such as fitting butt pads, an 8in (200mm) with coarse teeth on one face and fine on the other makes things fairly easy. There are also rifflers that are small rasps, ideal to get into tricky sections inside a stock.

The American gun industry is particularly adept at producing specialist tools made in a simple manner and usually at an economic price. Inletting and bottoming tools, scrapers of various shapes and sizes, are sometimes available for just a few dollars. They do not have the beauty of boxwood or rosewood handles, or the elegance of design of some of the more traditional and expensive tools, but they will do the job, usually very efficiently.

Specialist Tools
There are some tools that even the gunsmith who only intends to do basic servicing and minor repairs cannot do without.

Mainspring Vice A mainspring vice or clamp is used when dismantling and reassembling sidelock or hammergun locks fitted with the traditional vee spring. Rather than fighting the spring with a modified pair of pliers and risking damage to the mainspring, the vice spreads the load evenly, holds the spring firmly and give the user full control. With the continued growth of interest in muzzle loaders and early reproduction breech loaders, mainspring vices are still produced commercially.

Disc-Set Striker Remover Disc-set strikers can be exceptionally stubborn to remove, and a sturdy and snug-fitting tool is essential to avoid damage. They are available at prices that make it an uneconomic exercise to produce your own.

Dent Raiser A hydraulic dent raiser is a form of hydraulic jack designed to push out barrel dents. Those available at present are beautifully made and fairly expensive, and, of course, a different one is needed for each bore or gauge size (twelve, sixteen and twenty). It takes a fair amount of practice (best carried out on scrap barrels) and skill to use a hydraulic dent raiser, to avoid any possibility of bulging the barrel. However, when the technique is mastered, it is a particularly efficient method of dent removal.

If it is felt that the price is prohibitive, gunsmith-made mechanical dent raisers are a practical alternative, and available from a specialist supplier.

Bore Gauge A bore gauge is simply a comparator, but the most efficient method of shotgun bore measurement. For serious barrel work it is quite indispensable and far superior to a plug gauge.

Spring Setting Cramp The spring setting cramp is useful for the gunsmith who makes his own vee springs. It enables the operation of the spring to be examined when not fitted to the lock, particularly useful with Anson & Deeley actions where the spring is not visible in operation. It is not so necessary with sidelocks (external or internal hammer), where the full operation of the spring can be checked when assembled into the lock.

Snap Caps/Dummy Cartridges These are not strictly tools, but are necessary because a gun should not be fired on an empty chamber, requiring the cushioning blow of a cartridge primer to avoid broken or jammed firing pins. Snap caps are essential for workshop testing, and the best sort are those with replaceable inserts which take the place of the primer in a live cartridge. Once the inserts become deeply dented they no longer contact the firing pin, so the snap cap then serves no useful purpose.

Snap caps for testing pump action and self-loading shotgun functioning are made the full length of a cartridge and of a weight to simulate a loaded cartridge. Occasionally on the market are dummy cartridges which take a live primer and are very useful for checking firing pin strike. These are surprisingly noisy in a workshop environment and hearing protection is a necessity.

Making Your Own Tooling

Some regularly used tools are simply not available commercially. To the skilled gunsmith in the past this was never a problem, and many tools were made in the workshop, occasionally to accommodate a particularly tricky, once-in-a-lifetime job. Many still are – I have under my bench a box of tools and jigs, which have gradually grown in number and some of which have been modified more than once. The better tools are made from steel; a few, for convenience, are aluminium or brass. Those for one-off or occasional light use may even be made from hardwood. It is not uncommon for the following tools to be gunsmith-made for regular use.

Blacker or Smoke Lamp

The blacker is used to smoke-blacken parts when the finest of fitting is required. It is really a paraffin lamp with the wick left long, to produce the smoke. An additional, smaller lamp is an efficient means of lighting the blacker, rather than constantly using matches or a cigarette lighter. It is tempting to make one of these out of a small, screw-top glass jar so the level of fuel can be easily seen. However, this is not really advisable as it is potentially a simple miniature 'Molotov cocktail' sitting on the workbench!

Chamber Length Gauges

The length of chamber of a larger-bore shotgun can, with care, be checked with a 6in (15cm) steel rule. Smaller bores present more of a problem and, with any size, it is much easier to use a simple gauge turned on the lathe out of steel, brass or aluminium.

Top Lever Clamp

A piece of spring steel about ½ × ⅛in thick (12 × 3mm) with a couple or three slots of different sizes fulfils this role quite well.

Simple gunsmith-made chamber length gauge. Usually made from steel or brass, or even hardwood if nothing else is available. The flat on the handle is a convenient place to stamp the chamber size and also prevents it rolling off the bench.

Firing Pin Gauge

This is used to check the maximum protrusion of a firing pin against the breech face. A piece of brass or steel strip about ½ × ⅛in (12 × 1.5mm) with a

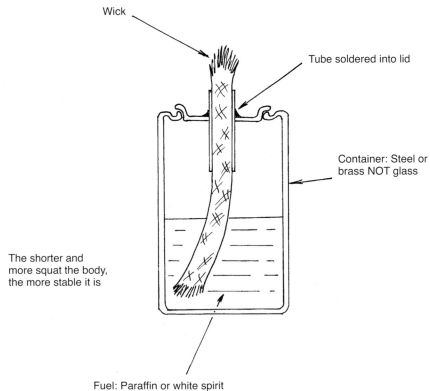

Wick

Tube soldered into lid

Container: Steel or brass NOT glass

The basic principles of a blacker – really a primitive lamp with the wick left too long – sometimes called a smoke lamp.

The shorter and more squat the body, the more stable it is

Fuel: Paraffin or white spirit NEVER petrol

round-nosed slot filed in one end makes a simple gauge.

Anson & Deeley Hammer/Mainspring Compressor

Almost all English boxlocks built on the Anson & Deeley principle, with either fixed or separate firing pins, require the mainspring to be in place and engaged on the hammer, and the assembly compressed, before the pivot pin can be fitted. No standardized tool seems to have evolved for what is a rather 'three-handed' job. A simple tool for this job can be made from steel, brass bar or hardwood.

Many Spanish boxlocks have the action slot, where the mainspring lies, opened out parallel so that the hammer can be put in place and the mainspring fitted through the front of the action and pressed into place with a simple tool comprising little more than a piece of brass in a file handle. It is a much more efficient way of installing a mainspring and it is hard to understand why British gunmakers, with the modern machine tools now available, do not generally adopt this simple idea.

Dismantling Block

When an action is being stripped it needs to be supported, which can be done against the vice soft jaws or the bench horse. However, when it comes to driving out pins or removing small screws, these functions are best done on the bench, as there is nothing more frustrating than scrabbling around for parts dropped on the floor. This is the time when a block with a soft facing makes a useful support. Blind holes cut into the block also make an ideal captive receptacle when driving out pins.

Wall Thickness Gauge

In the USA, there are proprietary wall thickness gauges, but all those I have ever seen in the UK are gunsmith-made. They are usually made to check the thickness up to about 10in (250mm) from the muzzle. This more than covers the choke area and the thinnest barrel section just before the choke.

Stock Clamp

Sometimes when repairs have been done around the head of a stock it is necessary to hold the stock on to the action without fitting the screws (pins), or even the trigger plate. The stock clamp is simple, essential and easily made from a couple of pieces of wood and two lengths of screwed rod.

Heat Sinks and Barrel Irons

Similar items but for different purposes. A heat sink

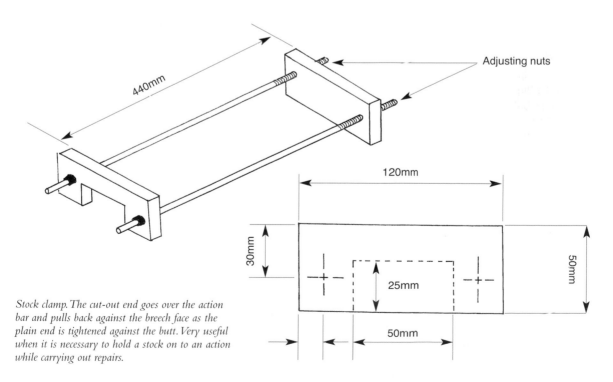

Stock clamp. The cut-out end goes over the action bar and pulls back against the breech face as the plain end is tightened against the butt. Very useful when it is necessary to hold a stock on to an action while carrying out repairs.

440mm

Adjusting nuts

120mm

30mm

25mm

50mm

50mm

is used to draw heat away from an area; an iron to apply heat. The heat sinks I use are steel with brass handles; the irons are brass with steel handles. The reason for this reversal of construction becomes obvious in use.

Polishing Formers

Hand polishing, particularly of double barrels, where ends of barrels and edges of ribs should be neat and well defined, is not achievable without some support for the fine abrasive or polishing paper used. A former, for carrying out the bulk of polishing on uninterrupted lengths of barrels, is easily made by cutting a half-round section into a cork block of the type used by cabinetmakers. To get into the area where barrels and ribs join, a thin vee of hardwood makes a good backing support. When it is vital to produce a flat surface, the job can be done with an old file with the teeth partly ground down to provide a grip on the abrasive paper used, but not to pierce through. For curved surfaces, such as the top of some top ribs and most bottom ribs, hardwood formers of different diameters can be used.

Chequering Tools

Chequering tools are available commercially, but, as much of a gunsmith's work involving chequering will be recutting or cleaning, it usually means matching a tool to existing line spacing, so it often becomes necessary to make your own.

Chequering tools can easily be made from a piece of spring steel and a file handle, with either a single row of teeth, or double teeth cut with needle files. I harden and temper my chequering tools, but some gunsmiths leave them soft on the basis that, although requiring frequent sharpening, they are adequate to do the work and are easily re-formed if required for a different job.

Screw Brace

The really awkward screw that appears in good condition but proves impossible to move with a conventional screwdriver ('frozen' is the term used to describe this condition) often needs just a little more leverage.

Adapting a woodworking brace to use with short screw bits is quite easy and many of the older gunmakers use a similar but heavier-duty tool. The screw bits can be modified proprietary hexagon bits, which can be held in a brace, although not a perfect fit, or the shanks of old worn-out wood drill bits cut off and reshaped with screwdriver ends.

Sear Stoning Jig

Many woodworking tool catalogues list a jig for holding a chisel while stoning it to produce a consistent shape. An extension of this idea is to make a jib that will hold a variety of gun sears so that they can be trued or the angles corrected to get the required trigger pull.

Mechanical Aids

The Lathe

No single piece of mechanical equipment is as useful as the lathe. It can be used as a miller, a borer and for lapping and polishing, as well as for the normal production of turned parts and tools. Virtually nothing is impossible with a lathe as long as it has all the correct accessories and the user is prepared to accept that certain functions are more efficiently carried out with specialist equipment. In other words, the setting-up time may be a factor that militates against its use for certain jobs.

A lathe is better larger than smaller. Small lathes are of limited use for boring, lapping and polishing, and these 'hobbyist'-type machines always fetch a premium price. A lathe of around 26in (660mm) between centres, with a 4½in (115mm) swing — swing is the measurement from the lathe bed to the centre of the chuck — is a useful minimum size and usually cheaper to buy second-hand.

With an older lathe simplicity is a desirable feature, as spares may no longer be available. Without any doubt, one of the best lathes for the gunsmith's use is the belt-driven American-made Southbend. Sturdy, quiet and comparatively simple in construction, it is everything the gunsmith needs. Although the design is now fairly elderly, examples do occasionally come up for sale in reasonably good condition.

Whatever the lathe, it is of very limited use without accessories and tooling. Three- and four-jaw chucks, running and fixed centres, a tailstock chuck and at least a fixed steady — a travelling steady is a bonus — are all essential equipment. For older lathes without a screw cutting gearbox, a complete set of change wheels is of vital importance, as it is not uncommon to find an important and frequently used change wheel missing.

The best advice I can offer concerning lathes and their use is to find a copy of *The Amateur's Lathe* by L. Sparrey. Occasionally reprinted, it is a superb example of a concise, no-nonsense approach to the lathe and its functions.

Drill Press

The drill press is, next to the lathe, one of the most frequently used pieces of equipment. A small bench-mounted vertical mill/drill is a useful substitute if a small milling facility is required but space at a premium.

Miller

There are two types of miller: the vertical and the horizontal. Some, such as the well-known Bridgeport, are convertible for either use. For general-purpose use, the vertical miller is the most useful and saves a lot of file work.

To the general gunsmith, the horizontal miller has limited appeal; deactivation of firearms is a modest industry, and this is one area where it can be better than the vertical miller. As with the lathe, there is a requirement for accessories and tooling, including a machine vice or a set of clamps to hold work. Cutters, of course, should be of the correct type to fit the tool holder.

Tool Grinder

A grinder is needed to sharpen and touch up lathe tools and the like. A small double-ended grinder with 6 or 8in (150 or 200mm) wheels is adequate.

Horizontal Grinder

Rarely seen in a gunsmith's workshop, this is an important, if not often used, piece of equipment. It is mainly used for work on trigger mechanisms, a function that cannot be equalled by hand. The small type, often described as a tool-room model, is more than adequate and a magnetic vice is a most desirable time-saving accessory.

Sanding Disc

For some aspects of stock work, such as fitting butt pads and extensions, a sanding disc of around 80-grade grit saves vast amounts of time and produces accurate work. Discs are available commercially and also as part of a small sander/linisher machine. A disc can be set up on a lathe, but it is just as easy and more convenient to make a sander with a wooden base, aluminium disc and a washing-machine motor. For stock work a disc needs to be 10 to 12in (255 to 305mm) diameter.

Linisher

I find less use for the linisher than for the sanding disc, but it has its uses for fine work, and belts can be changed relatively easily. A sander that takes a 2in (50mm) belt is a useful size; anything smaller should be avoided.

Polisher

A polisher can be purchased, adapted from a double-ended grinder or made in the workshop. Whatever the chosen method, it is best to have a wheel or mop that rotates away from the work, particularly when dealing with long components such as barrels.

Spill Borer

Once in common use in the gun trade, the spill borer is now rarely used; to a great extent, it has been replaced by the honing machine. It does still have some advantages for certain barrel work and, again, can be set up on a lathe or using a dedicated machine.

Lap

If any serious barrel work is to be considered, or even choke alterations, then a lap is the simplest tool for finishing. Lead laps can be cast in the workshop and are often used set up in the lathe.

Hone

The hone is, in simple terms, the sophisticated form of the lap. It can be adjusted mechanically and a variety of stones can easily be used. A hone can be set up like the lap, in a lathe or using a proprietary honing machine.

Ultrasonic Bath

I was introduced to this unusual but useful piece of equipment by our local clock repairer, who, rather than completely stripping a clock mechanism, cleans it in the ultrasonic bath before lubricating it. The same principle can be applied to gun locks. It is not suitable for all gun locks; some, like boxlocks, require complete stripping out, while side locks are particularly easy to deal with on the bench. However, for the trigger assemblies out of pump-action and self-loading shotguns and some over-and-unders, it does a very good job, and you can get on with something else while cleaning is in progress. An ultrasonic bath about $5 \times 8 \times 3$in deep ($120 \times 200 \times 90$mm) is sufficient for lockwork. When they do come up for sale second-hand, they are often very cheap.

Gas Torch

A propane torch with several sizes of nozzle is necessary for many soft-soldering and silver-soldering operations, as well as simple case hardening.

Oxyacetylene

A small oxyacetylene set is superior to propane

when precise silver-soldering of a small component is required and is also, of course, necessary for welding. For building up worn sears it is essential equipment (although the same job can be done with MIG and TIG welding).

Notes on Mechanical Equipment

When mechanical equipment is used by anyone other than the owner, there are serious Health and Safety considerations in the UK. For example, many machines now have to be fitted with brakes that stop the rotating parts within a certain time once the power is shut off. Older machines, and even the simplest 'home-bred' machine, may not comply with all the latest legislation, particularly if employees have access to that machinery, no matter how limited. However, at the moment the self-employed gunsmith, working unaccompanied or with sole access to machinery, is subject to fewer restrictions.

Many 'accidents' are avoidable. Exposed belts should be guarded, and safety glasses and gloves must be worn when applicable. Stop buttons should be easily to hand, and preferably either of the two-button type (red for off, green for on, with the red button prominent), or the single button pull for on, push for off sort.

Loose clothing, ties and personal jewellery, such as rings, should always be absent if you wish to avoid possible strangulation (semi- or terminal) or ripping a finger off. I once asked an old gunsmith why he had such ancient machinery with exposed flat belt drives. Flat belts, he told me, are better – they only bruise your fingers or maybe break them. Modern vee belts chop them off, he said, holding up the stumps of two fingers to make his point. Make sure you take precautions to avoid either consequence – bruised or missing.

CHAPTER 3

Examination for Faults

Introduction

A worn-out shotgun is just that – no more, no less – possibly non-serviceable, even dangerous, maybe beyond economic repair. The best approach to examining any gun is to assume it is riddled with faults and has been bodged unmercifully over the years, then you will never be disappointed.

It is surprising how many people (including some who should know better) can be seduced by a famous name or an action of particular technical interest. Add some good (albeit often slightly worn) engraving, handsome walnut and perhaps a fitted gun case complete with trade label, and a prospective purchaser will go weak at the knees as their common sense evaporates.

It is a simple and undeniable fact that anything mechanical will, with sufficient use, eventually wear out. Add a little abuse and negligence, perhaps a lack of maintenance, and the process is accelerated. Due to their inherent quality the best-made guns will stand long usage better than most mass-produced items, yet if they are not meticulously maintained, they too will eventually develop faults. Some mass-produced guns are also capable of good service. One Beretta over-and-under came into my workshop after firing over 30,000 cartridges in three years while out in all weathers. Apart from cleaning the bores and an external wipe-over, it had received no maintenance, yet its only repairs were a replacement bottom barrel extractor and top firing pin.

There are a few other guns that shoot loose with only the slightest provocation, have top levers that will not stay open and ejector mechanisms that display an independence of operation that is sometimes alarming. The worst of these guns give the impression that the mechanism is constructed of slightly hardened putty! They are a bitter disappointment to the owner and a disgrace to their makers, who often do not appear to understand even the basic principles of simplicity and reliability. Even the much-derided cheap hammer guns sold to the impecunious in the period between the two World Wars were superior to some of the flashy junk on the modern market.

The Name

Corruption and Confusion

A name that seems at first glance so familiar may not be all it seems. It is occasionally possible to find guns with what appear to be variations on well-known names or similar names to famous makers. The heyday of this practice was during the latter period of hammer-gun production when the Birmingham trade turned out thousands for retailers. It was not unusual for a local shop to have guns bearing their name, often engraved within a panel on the top rib just in front of the breech face. So a double hammer gun bearing Birmingham proof marks and the legend 'Osborne – Evesham' is not a hitherto unknown

Originally listed as gunmakers, starting in 1853, Osborne's traded as ironmongers and it seems certain that most, if not all, of the guns were actually made in Birmingham, a not unusual arrangement.

gunmaker from a Worcestershire market town much better known for the production of plums and asparagus, but, in fact, one of the local ironmongers, albeit one that advertised as a gunmaker.

Fifty years ago there were still many old guns in use and it was possible to find all sorts of names. Some were company names, others created to associate the product with the cachet of 'best' quality, a rare few simply a kind of forgery intended to deceive the unwary. Some still surface even now but, with the passage of time and the disappearance of so many small-time gunsmiths who would make the occasional gun, and retailers using their name or a brand name, it can be difficult to trace the real origin. However, 'Goggeswell and Harrison', even allowing for an optimistic eye and a little wear on the engraving, is not Cogswell & Harrison. 'Purdy' is spelt with an 'e' if it is indeed a product of the company now known as James Purdey and Sons.

The corruption or mispronunciation of names does not help with either identification or research. The fine-quality Birmingham gunmakers, Westley Richards & Co. is not 'Wesley', 'Welesley' or 'W. Richards'. The latter may be the fairly well-known Liverpool maker but there may also be found 'W. Richard', which gives no clue to the place of origin apart from the ubiquitous Birmingham proof marks. 'Holland and Holland Ltd. Residing at 31-33 Bruton Street, London' is not Holland and Hollands, although this inaccuracy may have come about via the not-uncommon practice of referring in speech to the company simply as 'Hollands'. 'H. Holland', though, may be Harris Holland, a founder of this famous firm.

With much older guns, usually muzzle loaders, confusion may arise simply through the English spellings in use during a particular period. For instance, 'J. Blisset' and 'I. Blisset' may be the same maker as the letter J and I were, for a time, interchangeable. It was not unknown for ribs to be swapped, not with any intention to deceive, but simply to replace a damaged or corroded rib. One gunmaker I used to visit had a small barrel full of second-hand ribs for just such a purpose. With a quick economy job sometimes the name on the rib would not be struck off so either it did not match the name on the locks or, if they carried no name, the gun simply assumed a new identity!

Researching a Name

Names do change, companies alter and gunmakers move workshops, so different addresses will appear following a maker's name and this can be a pointer to the period of production. A surprising number

At one time I had a William Powell boxlock non-ejector pass through the workshop. It was a particularly handsome Damascus barrelled gun with scalloped back to the action, very fine engraving and good walnut. On contacting Powell's (who are unfailingly helpful), I was informed that it was some fifteen years older than I had 'guesstimated', built on a best action that would normally have been an ejector gun. The customer must have wanted a non-ejector, and the gun was fitted some years later with disc-set strikers.

of gunmakers have kept their old records and the easiest way to validate a gun's origin is to check with them. It not only adds provenance but sometimes value.

A window into the past can add considerable interest to a gun, but do not expect this service for free. Many makers will politely oblige with the year or approximate period of manufacture via telephone or e-mail, but copies of records or letters of authenticity cost them time and money. It is still usually worth it. The time you wish you had not bothered is when, say, comparison with the original records proves a gun has been rebarrelled with steel replacing Damascus and not by the original maker, or has had the barrels shortened to accommodate damage or fashion and perhaps been altered considerably in the stock. It is not unusual to find out that the particularly pretty little single-barrel .410 was once a rook rifle.

A primary mistake is to assume a letter of authenticity relates to the saleable condition of the gun. A letter may confirm only that it is, indeed, the product of a particular maker, and the date of manufacture. It may go further and give a full description, including original choke borings and stock dimensions. However, unless the letter contains specific information relating to condition or reference to a full examination and report, it is not an indication of the gun's soundness, only an historical reference.

Where a maker has ceased trading years ago research can be a problem. Some gunmakers were not the best record-keepers and, when a firm finished, papers were often thrown away with scant regard for history, perhaps assuming that they would never be of any interest to future generations.

There are many useful books as initial sources of reference detailing London and Birmingham makers, but there is much less information available on other makers. Nigel Brown, formerly Secretary of the Gun Trade Association, has produced the very interesting *British Gunmakers*, Volumes I and II, and the latter is especially useful for Birmingham, Scotland and the regions. These books also list names that are owned by other firms, some of whom may have the original records. Otherwise, there are records held by bodies such as the Royal Armouries Museum, Imperial War Museum and Birmingham Science Museum.

There may be some pleasant surprises in store when carrying out research, like finding out that a gunmaker that is thought to have disappeared still exists. Quite a number of the smaller British gunmakers seemed to gracefully expire in the 1960s and it is easy to assume, on reaching an apparent dead end, that a particular maker has closed its doors. Sometimes a little more persistence will reveal they have moved to a retail shop that is none the less a successor of the original makers. The Gun Trade Association can be of assistance and, as with all research, sometimes you might just get lucky!

General Examination

The idea of quickly checking over a gun and giving an opinion is attractive, limited and fraught with problems. Attend a game or clay pigeon shoot and as soon as word gets around that you are a gunsmith there will be a small queue of people who want to talk to you about work, when you are out just to enjoy yourself. Conversations revolve around all sorts of guns, from the one that someone intended to bring in at the end of the last shooting season for a service, to the old hammer gun 'that's now a bit loose' and reputed to have the longest range of any gun before or since.

I am always wary of the member who asks what I think of their newly acquired gun and, even worse, what it might be worth. Many men, in particular, seem keen for me to confirm to them that their new pride and joy was the bargain purchase of a lifetime. If, on a cursory examination, I were to list a catalogue of minor faults and give an estimate of value somewhat less than the owner had paid, he would be very likely to be offended, especially if this took place within earshot of his companions. My strategy is always to make suitable, if non-committal, admiring noises and to mention a broad price range that hopefully encompasses what he paid. I then go on to explain that a proper examination can only take place on the workbench, with the appropriate gauges to hand.

There are times when it is necessary to examine a gun without recourse to the full workshop facilities, such as at an auction. In the UK and many European countries gun barrels do have to be within their proof size to be offered for sale, although the degree to which they may be in proof can be marginal, and even at an auction I would take the bore gauges to check actual internal barrel sizes. It is also important to check the proof marks, whether nitro or black powder proof, and chamber length.

Whenever possible it is best to have the item on the bench and for side-by-sides, over-and-unders and single-barrel break-open shotguns the basic principles of examination are the same. Unless there is something obviously broken or missing, self-loading shotguns usually have to be test fired as their main fault is malfunctioning during the firing/reloading cycle. Pump-action guns, on the

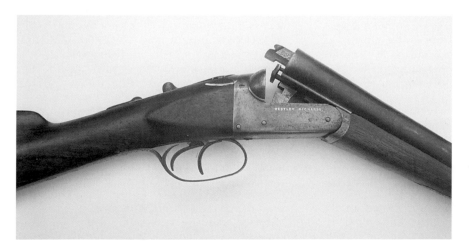

Westley Richards sixteen-bore with worn chequering, damaged top lever and rust-stained action, having seen considerable use. A so-called 'Gold Name' model, while one of the least expensive Westley guns it was none the less a credit to the maker. It had withstood the most appalling abuse, but it still worked!

other hand, as the functioning is hand-operated, can usually be adequately tested in the workshop using dummy cartridges.

Appearance

To the experienced eye, the general appearance of a gun can give a good clue as to its condition. If the chequering is almost worn away and the wood has 'shrunk' below the locks, the gun has, at the least, been carried and handled a lot, or may have had wood removed. Particularly dark staining in the walnut around the head of the stock indicates that someone has been in the habit of over-lubricating the action and locks. Given time, mineral oil will destroy the strength of even the best-seasoned walnut. Worn engraving is a good pointer to extended use but the odd dents on a stock could be just honourable scars obtained during a lifetime of otherwise cared-for service.

A dirty gun, and by that I mean a gun externally covered in a dark layer of usually greasy grime, may hold hidden horrors – or it may have been protected by that mucky layer for many years. Even in a grimy condition the sheer elegance of a finely made shotgun will show through, so you should not necessarily be put off by first appearances.

Side-by-Sides, Over-and-Unders and Break-Open Singles

To check the fit of the action on side-by-sides, over-and-unders and break-open singles, you should follow the same procedure:

1 After checking that the gun is clear (unloaded), close it and take note of the sound as the action shuts. Ideally it should snick together with a minimum of mechanical noise as the action locks shut. If it closes with a 'clacking' noise (a little like shutting a gate), the barrels are not locked tightly to the action.

2 Take note of the position of the top lever, the tail of which should, with the gun closed, be positioned just to the right of an imaginary centre-line down the top tang. If it is to the left then the locking bolt to which it is connected may not be holding the gun firmly shut due to one or more faults: wear on the locking bolt or bites where the bolt engages, or more often wear on the bolt and hinge pin allowing the barrels to move forward from the breech face so the bolt is partially disengaged.

3 With the gun still closed remove the forend and, holding the gun with one hand around the grip of the stock, with thumb and forefinger of the other hand on either side of the action, gently shake the gun from side to side. Particularly with a side-by-side, if there is any looseness it will be felt through the thumb and finger. In extreme cases of wear it can be heard and it is sometimes possible to move the barrels from side to side and up and down while watching the movement between the barrel breech ends and action face.

4 If, with the forend reattached, the barrels and action are not so loose, the gun may have been 'tightened' on the forend. While it makes the opening and closing action of the gun feel better it actually pushes the barrels away from the breech so that, when the gun is held up to the light, a gap can be seen between the barrels and breech face. With practice a gap of a few thousandths of an inch can be identified. It is less easy to detect with an over-and-under with high side walls but for those with shallower actions, side-by-sides and singles, it is a very good check even if the forend is unaltered.

5 For definite evidence of an over-tight forend, check whether, on opening the gun and pushing the barrels down to their furthest point of travel, the movement becomes progressively stiffer.

You may be wondering why this is so important, and the answer is quite simple: a gun should be checked against the standard necessary for submission to Proof House examination and testing. A gun has to stand the original proof test and any subsequent retest without the forend attached, and remain tight. A story has grown up that a certain famous model of over-and-under is only ever properly tight when the forend is locked in place. Usually that is just sales talk – it is certainly not how a second-hand gun should be.

Fixed-Barrel Guns

Fixed-barrel guns, such as bolt-actions, Martini-actions and oddities like the French-made Darne with its moving breech, do not suffer from the same wear patterns as break-barrel guns. Most of these guns have few problems with breech locking although the linkage on the Darne can become worn.

When excess headspace occurs with a fixed-barrel gun it results in fired cases displaying bulged rims, or proud primers after firing. The way to check these is to use headspace gauges or dummy cartridges with steel or brass shim added to measure the gap. Obviously, if there are doubts concerning headspace with pump actions or semi-autos, the same applies.

The Barrels

The barrels are truly the heart of any shotgun; if they are worn or damaged to a degree that renders them irreparable, then the gun is useless.

With the barrel(s) detached from the action, check for pits, dents and bulges, looking from both ends of the barrel(s). A set of bores that looks reasonably clean from the breech end may hold some nasty surprises when viewed from the muzzle.

Also, while looking down the muzzle ends of the barrel, examine the choke area carefully. When chokes are opened out they should, as the final part of the process, be lapped and polished or honed so that the surface finish is at least as good as the rest of the barrel. Chokes that have just been bored out and left as machined can look like a ploughed field and, as well as this being undesirable, it is only half a job.

Dents and bulges are the same type of deformation, the one appearing as an inverted image of the other. Dents are caused by knocking barrels against a hard object while bulges are caused by pressure on a thin section of barrel, brought about by an obstruction or over-enthusiastic use of the dent raiser. Both can be detected by looking down the length of a barrel viewed against a soft, constant light – if it is too bright, it can be confusing as well as hard on the eyes. Looking down the lines of light reflected along and inside the barrel will show deviations as they flow down into a dent or over a bulge. Some damage can appear as both where a dent has forced up a bulge of metal alongside.

Bulges are generally more concerning than dents, the latter appearing less often in the very strong modern steel barrels. Anyway, the cause of a dent is obvious while a bulge can be caused by something more sinister. Where a dent has been left unrepaired the passage of shot will, given sufficient use, wear it away to leave a thin section that may eventually erupt as a bulge. Over-enthusiastic striking down of the outside of barrels for refinishing or excess removal of metal in the bores will ultimately produce thin barrel walls just before the choke section. This can start to give way and will be seen as a long bulge or a kind of corrugation called rivelling.

A ring bulge is a complete circle, often of startling symmetry, which, even with a double-barrelled gun, continues under the ribs. The normal cause is an obstruction such as mud or snow that has entered via the muzzle, or a wad from a cartridge that has failed to fire properly. The latter is almost unknown with modern cartridges in good condition but thirty-odd years ago I had four cartridges out of a box of ten fail in this manner, leaving the wad in each instance stuck in the choke.

Pitting is corrosion and may actually be little more than light surface rust or deeper rusting that reduces the thickness of the barrel wall. Viewed, of necessity, mainly sideways on from the end of a barrel, even a modest patch of discolouration can appear dramatic, but with even a little experience this is discernible from the blacker, more dense look of proper pitting, which also has a sharper, more defined edge to it. One of the worst forms of corrosion occurs in Damascus barrels along the forge weld line; in the worst instances, it can be seen as a complete spiral of corrosion.

Pitting is the great unknown and large or deep-looking pits should be treated with the utmost suspicion. With a double-barrelled gun look for pits that occur on the sides, where the barrels are joined together. If a rib is loose then there is the likelihood of corrosion between the pair of barrels. In other words, a barrel wall may be being eaten away from both sides. This becomes an accident waiting to happen and one day someone's fingers, hand or eyesight may be lost as a result.

Look also for what is literally under your nose because even the chamber may be corroded, especially where cartridges shorter than the nominal chamber length have been used. Neither is it unusual with old guns to find a complete ring of corrosion in the forcing cone just in front of the chamber at the start of the bore. Fortunately, the more modern hard chromed bores do not suffer such corrosion problems.

Before leaving the subject of barrel bores, there are a couple of other visual checks to do. Even if the bores are clean, look at the lines of reflected light to make sure they are straight. Curved lines mean the barrel is not straight – indeed, it may be bent as a banana. While you are at this, focus through one barrel at an object about thirty to forty paces distant (I use the house television aerial), then switch the eye to the other barrel while holding the barrels perfectly still; this should give the same sight picture. If it does not – and it is worth a double check at this point – the barrels have been laid without them firing to the same point of aim. This can be verified later by firing at a pattern plate, sand bank or, at the right time of year, a snow drift.

Fine concentric circles up a bore indicate that it has been re-bored, something which can usually be confirmed when bore gauging. 'Wobbly' lines of light down the full length of a barrel, however, might mean you are actually looking down the bore of an oval-bored rifle. Nineteenth-century Lancaster rifles appeared in sizes such as sixteen- and fourteen-bore.

With gun barrels size is important, particularly in countries that have proof laws governing the condition for sale. For a thorough examination the best tool is, without any doubt, the bore comparator. It can be used to check not only the bore size at the datum of 9in (23cm) from the breech end of the barrels, but also other dimensions. Older barrels often have tapered bores. This is usually a by-product of spill boring and not a disadvantage, but you should check either side of the datum. It is rare, but not unknown, to find the bore has been honed either side of the datum to clean up blemishes, leaving the barrels on their proof size at only the 'magic 9-in mark', and producing an undulating bore, which is a very doubtful practice.

The comparator can be used to check the choke sizes relating to the bores and the shape of the chokes. Choke sections should definitely not be bell-mouthed.

If a wall-thickness gauge is not available, a guide to wall thickness at the most vulnerable area, just before the choke, can be found by measuring the inside dimensions with the bore comparator and the outside with a micrometer, and then using the following simple formula:

$$\text{Outside diameter minus bore size} = \frac{\text{total thickness}}{2} = \text{individual thickness}$$

This is only a guide – when barrels are struck up, the thickness either side may vary. Also, old barrels often have the thinnest area on the outside, which, being opposite the rib with doubles, is impossible to measure by this method. The result will not be exact, but if thicknesses of around 0.030 to 0.035in (0.90mm) are being recorded, that is reasonably reliable. Minimum wall thicknesses of, say, 0.023in (0.60mm) in good barrels with modest choke will often go through proof again. The EU has, in the past, proposed a mandatory minimum wall thickness regardless of whether a gun goes through

The bore comparator is, without doubt, the best instrument for gauging barrel bores.

proof. If this ever became law, the nature of bureaucracy suggests such a minimum thickness would be very much on the 'safe' side. At a stroke of the pen, many fine British guns otherwise in proof could become unsaleable.

Obviously, wall thickness is best checked with the appropriate gauge and the gunsmith-made 'post type', while not 100 per cent reliable, if used to record only the minimum readings will be on the safe side. The proprietary calliper comparators with a small bearing area are possibly a little more accurate and it is quite easy to build up a profile of barrel wall thickness for as far as the callipers will reach. Again, this cannot be done with a double-barrelled gun in the area masked by the ribs and the other barrel. However, this is usually the thickest area, being well protected and not offering the possibility to 'strike up' for any refinishing.

The Ribs
When barrel ribs become loose – meaning continuous, not ventilated ribs, as found joining the barrels of conventional side-by-sides and many over-and-unders – hidden rusting can take place. Even with a slight gap between rib and barrel, water will enter by capillary action, particularly with the top rib of a side-by-side, which is most exposed to the elements when out in the field.

Where a rib has been loose for a considerable amount of time, with the gun remaining in service, it is not unusual on removing the ribs to find rust flakes packed between the barrels. In the most extreme circumstances, given enough time, such a build-up of rusting will actually exert pressure against the ribs. Add to that the corrosion that can occur along the joint line of rib and barrel, and the

rib can be rotten to a degree that requires complete replacement. With some unusual rib forms, this may prove very difficult. Not only that, but the barrels between the ribs will also be eaten away, and possibly weakened to the point of being dangerous. A gun with loose ribs should always be treated with suspicion.

The problem can be compounded by the type of flux used when laying the ribs. At one time, elderly German-made combination guns or drillings had an unusually high failure rate in later years during British re-proof. They were often subjected to re-proof because for a long time following the Second World War, German proof marks were not accepted in the UK, and re-proof was necessary for a legitimate sale. The problem was allegedly due to the long-term corrosive effect of the flux used on assembly. Even with repair work in later years, care does need to be taken to ensure any traces of corrosive flux are cleaned away.

> I was once shown a remarkable example of failure in a double-barrelled sixteen-bore shotgun with a 8 x 57 JRS (8mm Mauser rimmed cartridge) barrel underneath. The rifle barrel had ruptured about halfway along its length and the bullet had torn its way out at a shallow angle between the shot barrels and through the top rib. Unfortunately the barrels were thrown away before a detailed and proper examination could be made to determine the possible fault, which may have had nothing at all to do with the methods of assembly used at the time of manufacture.

Rotten barrels – not a very technical term but an apt description of the type of corrosion that may be found under a loose rib.

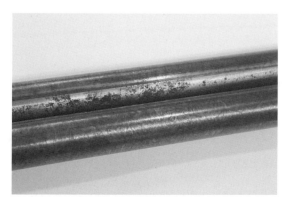

Many Spanish guns and some others originating from continental Europe have ribs silver-brazed in place, what is commonly called silver-soldered. On economy guns this is often visible as a gold-coloured line along the edge of the ribs where they contact the barrels and shows a fillet of material that has been left in place. It is sometimes assumed that this method of assembly is detrimental to laying a rib, as many appear as buckled as railway lines on a hot summer's day. Yet it is possible to find barrels assembled by this method where they are indistinguishable from those of soldered assembly, beautifully laid and with no joining medium visible. According to one of the world's best barrel makers it is more difficult to get really good results with the brazed form of assembly, although in its favour is the fact that they rarely come loose.

The test for sound and secure ribs is delightfully simple. Suspend the barrels by one finger from a convenient point such as the hook, and flick them with a fingernail of the other hand up and down their length. Fine-quality, well-laid barrels in good order will ring like tubular bells with a lingering resonance that is literally music to the ears. Heavy, thick-walled barrels emit a duller tone and sometimes need a good tap with the knuckle to respond. This is not necessarily confined to cheaper guns. Even with guns bearing reasonably well-known names there were periods when barrel steel seemed not at its best; or they may have been rebarrelled with tubes inferior to the originals.

If any part of the rib is not securely attached, either the barrels will not ring when tapped adjacent to the defective area, or a high-pitched vibrating rattle is the result. The latter denotes a really loose rib and it takes only a little practice to discern good barrels with soundly attached ribs. With a little more practice it is possible to detect a faulty joint to within an inch or two. Then it can be a case of probing with a slim, finely pointed scriber to see if the rib will lift.

What about the touchy subject of refinishing? Originality can be everything with collectable firearms but refinishing, such as blacking or browning of shotgun barrels, is generally accepted as not harmful to the value if done properly. A pair of nicely rebrowned Damascus twist barrels is a joy to behold and the work will usually increase the value. This is quite the opposite to polishing off an action and removing the remains of the colour case hardening to produce an 'old English finish'. Neither is refinishing of any sort acceptable to collectors of rifles and pistols.

The Lumps

The condition of the lumps tells quite a lot about the care and maintenance history of a gun. While part of the barrel assembly, they are the other half of the action of locking a gun soundly shut and potentially an area of the most unseemly 'repairs' usually carried out for short-term action tightening.

Ideally the lumps should be clean and square-cornered without any sign of bruising or welding, hammer or file marks. There are various methods of tightening or putting barrels back on the face, many of which involve building up the hook to refit to the hinge pin. These methods are discussed in Chapter 7. Other less orthodox, or shorter-term repairs are nothing more than vandalism, many carried out with a hammer, sometimes aided by

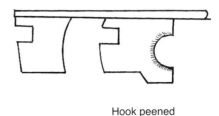

Hook peened

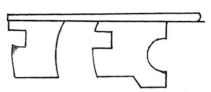

Rear bite peened

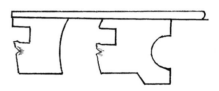

Bites chiselled upwards and filed

Some of the bodgesmith's methods of tightening. From top to bottom: sides of hook peened, rear bite peened upwards, bites struck with cold chisel to bulge metal upwards.

the additional use of a centre-punch or chisel. Hammer marks around the side of the hook may be evidence of such work. Other more extreme methods are easily visible. Similar work might be found to the lower edge of the bite (usually the rear bite only where there are double bites). These should appear as a straight taper; a rounded-off form is a sign of wear. If the lower back edge of the bite is pushed sharply upwards, this is evidence of an attempt – usually by hitting with a hammer – to lock the barrels more firmly into the action. A somewhat more permanent form of this method involves a larger section of the bite being pushed upwards using a cold chisel. The resulting chisel cut is welded up and the bite re-cut to better engage the locking bolt.

The Action

The action should be examined externally for visible evidence of wear in the form of rubbing at the edges, worn engraving, and loss of finish or colour case hardening. The breech faces of old guns used with cartridges containing corrosive primers may be pitted but, as long as this is not excessive, it is unsightly rather than unsound. Scouring down the breech face of an over-and-under, particularly the type with constantly sprung extractors, or localized damage almost showing an imprint of the extractor may suggest that a new extractor has been fitted with the barrels off the face. The flats of a side-by-side should be exactly that: flat with no raised edges or curved faces. This is another area where the 'bodgesmith' method of tightening is sometimes found. Hammer marks along the edge of the flats to bulge metal upwards to fill in a gap between barrel and breech flats are not unknown; if carefully mixed with some borderline engraving, they can be cunningly disguised. Another unofficial method is to squeeze the action in the vice, which can push the internal sides of the action against the lumps as

well as bowing the flats. One method of checking the action flats is with a straight edge.

The Forend Iron

This, in its simplest terms, stops the gun falling apart on opening. It carries the forend wood and usually the ejector mechanism, if fitted. It is used in con-

---

I have seen an early Scottish breech loader with Jones rotary underlever, which at first gave the impression of a tight and tidy action. Close examination by holding it up to the light showed the barrels to be firmly against the face but revealed a large gap between the barrel flats and action. The action bar had been bent, having the effect of shortening the distance between the hinge pin and breech face, thereby pushing the barrels back against the standing breech. At the same time it moved the lock of the underlever further away, effectively taking up any wear on the bites and rotary locking lugs. To the untutored eye, it was a devious and cunning method of tightening that would be virtually undetectable. However, on this gun the perpetrator had overdone the action bending so the breech ends of the barrel had been filed to enable the gun to close but leaving the cartridge rim recesses too shallow. Whoever carried out this particular work had a reasonable knowledge of how a gun fits together but what they really succeeded in doing was damaging what had once been a fine and interesting early gun. Practitioners of this dark form of gunsmithing should be banished to the gunsmith's version of hell, a very junior clerical job in a government department in inner London.

THE BODGESMITH

Where there is evidence of a crude method of repair, it is unlikely that the work will have been carried out by a gunsmith, or at least not by anyone who wanted to stay in business. It is rare nowadays to encounter such unorthodox repairs, but in the 1950s the local blacksmith or car repairman was often called upon to effect some repair. Money was tight, transport limited and they had the basic essentials: mechanical knowledge, a large vice and a hammer. Some repairs were more sophisticated than others, although the sight of a patch brazed on to the outside of a holed barrel does not fill the shooter with confidence. Most often there would be a squeeze in the vice and perhaps a tap with the hammer and, in return for the favour, perhaps a brace of rabbits or a pheasant poached from the nearest sporting estate.

Now that they have more disposable income, few, if any, shooters need to scour hedgerows with an old, pitted and worn Birmingham or Belgian hammer gun, probably black-powder proofed but unwittingly loaded with nitro cartridges. Times have changed and shooters are more technically aware, which is often the first step on the road to caution.

Occasionally, though, economy repairs are practised by the gunsmith. What should he do when old Fred brings in his faithful Stevens single-barrel, bearing that wonderfully evocative address 'Chickopee Falls'? The barrel is massively strong, so strong even ancient examples are often still well in proof (although some reached customers in the UK by a variety of means and never saw the proof houses). The gun is next to worthless but old Fred does not want to change, neither does he want to spend a fortune on repairs and, anyway, the gun is not worth it. How the gunsmith tackles this is a matter of conscience. The only certainty is that whatever is done has to be safe, and the gunsmith is judged not by all the good jobs he has done, but by the one that fails or does not come up to an expected standard.

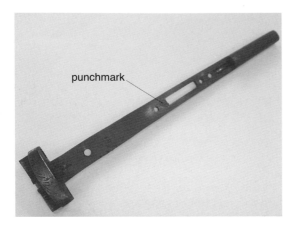

punchmark

A tightened forend iron. Overdone, this is a recipe for making a gun seem tight while it is actually off the face.

junction with opening the barrels to cock the locks of most hammerless guns and is fitted with some form of catch or cross bolt to hold it in place. There is very little to go wrong with this part unless it is lost, which renders the gun fairly useless but does make a very effective method of security!

The necessary opening and closing of a gun can cause wear at the knuckle with some galling if it was not well fitted in the first place. With spring-loaded snap-on forends such as the Hackett type, wear takes place at the back of the barrel loop and, to a lesser extent, in the corresponding mating area in the forend iron. If the forend has been tightened there may be hammer marks at the rear of the loop and metal bulged backwards to provide a tighter fit. If so, this again is the work of the bodgesmith. However, tightening by using a shaped punch to re-form the rear of the slot in the forend iron is generally acceptable. It is not acceptable when it is done to give the impression of a tight action.

The Ejectors

Ejectors can, when worn, be a bag of trouble and the potential for wear in some of the early types is quite alarming. The best type, as is often the case, is the simplest, and with side-by-sides that is the Southgate mechanism. This is now in almost universal use with English and continental side-by-sides.

The Greener system, using helical springs to power the kickers, is relatively trouble-free and straightforward to repair, as is the Westley Richards ejector box, which was supplied as a proprietary item to other gunmakers. If the latter has a fault compared to the Southgate system it is that the vee springs are shorter (vee springs work and last bet-

ter if on the long side) and, with hooked ends, more complicated to make.

Well-made and popular over-and-unders such as Miroku, Browning, Winchester and Beretta rarely suffer from ejector problems. The exception is the occasional broken extractor, which, when it lets go, is capable of ejecting the broken part into the bushes, never to be seen again. The Winchester based on the 101 action may, occasionally, be found with a broken ejector kicker (or hammer) but new heavy-duty replacements designed to overcome this problem are available. The once-popular Baikal range had a model where the ejectors could be set to act in non-ejector mode. It seemed an unnecessary complication and was sometimes a nuisance to the extent that it was not unusual to find guns always set as non-ejectors.

Testing ejectors should be done with snap caps, not only to ease the firing pin blow but to watch the actual action of ejection. Lightweight snap caps that nearly simulate the fired cartridge case weight are best. After firing both barrels, which with inertia-weight single-trigger guns will require a tap on the butt to simulate recoil, open slowly. It is easy to open a gun quickly and see an ejection that seems even, but by opening very slowly the ejectors can be heard to kick in. It is surprising how often there is a slight delay between the two. Mistimed ejectors are not a problem but it is something that should be attended to before a gun is offered for re-sale.

It is also worth operating each barrel independently to make sure they cleanly eject on their own.

The Locks

With hand-detachable side locks it is not a problem to examine the whole mechanism quickly, but normally no such luxury is afforded, so it is in the first instance a matter of external visual examination. With the hammers down in the fired position (which can be done by 'firing' the mechanism against a block of hardwood or the edge of a wooden bench), examine the end of the strikers or firing pin(s). Over-and-under firing pins seem to have a particular fondness for breaking and the small-diameter parallel end sometimes falls out of the firing pin hole, usually to disappear. At least the absence of a visible pin makes a misfire obvious.

Examine the ends of the strikers for wear, pitting and deformation, which are potential sources of trouble. Protrusion from the breech face can be checked with a gauge, but with rebounding locks it is necessary to push them forward manually to check. Except with hammer guns, this means removing the locks or stock. If you are this far into the job then a problem has already been identified.

An indication of the force behind each lock can, with practice, be identified by the sound of the locks snapped against a snap cap. One method favoured by some of the 'old boys' was to place a small coin over each firing pin hole with the stock/action held vertically, and then to pull the trigger(s). If the coin hit the workshop ceiling, it was deemed OK. If it merely leapt only a foot or two (about half a metre), then it was suspect.

This method does have some merits but I have never liked dry firing or snapping locks without something of reasonable substance to cushion the firing pin and hammer blow. This is where the snap cap made to take a primer is most useful. The indent of the strike is visible, although it will not leave quite the same impression as a loaded cartridge, due to the lack of breech pressure upon firing.

There is a difference between the conventional sidelock, descendant of the hammer-gun bar lock, and cheaper, simpler locks of similar appearance (*see* Chapter 1). The visually complex and expensive 'proper' sidelock has five, sometimes six, or even seven pins showing through the lock plate, while the simpler locks have only two or three. It is worth taking off a lock for a peek inside as sometimes dummy pins were retro-fitted to the cheaper locks to give the appearance of a conventional sidelock, thereby deceiving the unwary and increasing the potential sale value.

A three-pin Baker lock retro-fitted with four extra dummy pins, to give the impression of a more complex lock. To the initiated the shape of the lock plate gives it away, along with the later engraving and mock gold inlay of the name.

Stock and Forend

Wood is obviously more vulnerable to wear and neglect than steel. Apart from the obvious staining of excess mineral oil, you should also look for worn chequering, wood shrunk below the mating steel parts, cracks, missing wood and modifications. A few small dents are of no great consequence, except where they have broken the grain, or where it is not a dent but a tear that may have been produced by contact with barbed wire, for example.

The front of the forend where it wraps around the barrels on a side-by-side is vulnerable to damage. Check for reshaping that is sometimes done to accommodate missing wood. With an over-and-under, examine carefully for cracking at the rear end adjacent to the ejector kickers and underneath behind the forend catch where the wood is usually thinnest.

A shortened stock is something that can be overcome but if it is too short then any extension work will look odd and reduce the potential value. One way around this is to fit a plain piece of wood and then ink in the grain pattern; this is a method to which cabinetmakers sometimes resort. It is time-consuming, and therefore expensive, and might be regarded, in spite of the high degree of skill involved, as a form of cheating. As a quick check the average length of pull is about 14¼in (36cm).

Poorly fitted butt pads or plastic butt plates on a gun that would once have had horn or steel are undesirable additions.

Rechequering on an old stock is always worth careful examination. A magnifying glass is useful to detect repaired and over-chequered cracks or even dowels, which, with exposed end grain, will usually show up if turned in the light. However, sometimes a freshly broken stock that was once a jigsaw puzzle of parts can, in the hands of an expert, be invisibly repaired and only an internal examination will detect this.

Stocks can, like actions, develop looseness. This is of particular concern, as the more it is used the more the damage accelerates, until the stock may deteriorate into scrap. With an over-and-under of the stock bolt-type fixing, looseness usually occurs when the bolt has not been tightened properly. (But note that at one time the Russian Baikal of this type would sometimes display splits at the head of the stock along the grain, which appeared to be a fault of the wood, aggravated by the design.) On self-loading shotguns and pump-actions, the stocks are so simple and substantial that such faults rarely occur. A quality handmade English side-by-side (or, increasingly, over-and-under) with a loose stock can be a financial bad dream. Very early breech loaders, surprisingly, are not necessarily a problem. Many, such as the wood bar type, are of fairly substantial dimensions, very like a muzzle loader. Most problems are based around guns of a little later vintage, dating from at time when it became fashionable to use a short top strap with only one pin (the breech pin) through the stock as the primary means of holding everything together. As an expression of the gunmaker's art where the integrity of the fit keeps everything sound it is a superb piece of work, but as a means of mechanical soundness it is flawed. Even side-by-side stocks with long straps held together by two pins will come loose if too much mineral oil seeps into and weakens the stock, or if a split develops along the grain. The other problem is when stock pins – especially the action pin – are over-tightened, crushing the mating wood and compromising the proper fit of metal to wood. This is noticeable when the screw slots no longer line up along the axis of the gun.

To check stock tightness, hold the barrels and forend under one arm while attempting with the other hand to push the stock both up and down and side to side. Do not overdo this test. If it has been broken and poorly repaired, it might break!

Considerations for Repair

Guns will display a vast range of faults, damage and bodges, particularly the more elderly examples. In theory no repair is impossible, but it may be impractical or financially unrewarding. For example, a cheap mass-produced single-barrelled gun that has shot loose may not be economic to repair but the same judgement would not necessarily apply to an elegant English hammer underlever single of the 1880s. At the same time such a gun would not normally be considered for rebarrelling while a best-quality double would be a better financial proposition.

A few customers will have a gun repaired whatever the cost because of its sentimental value, but they may go on to regret allowing their heart to rule their head. It is the same approach as those people who spend enormous amounts restoring vintage cars or motorcycles to better than new condition, seemingly happy in the knowledge that what they are spending far exceeds the market value of the vehicle.

The two areas of major cost in gun restoration are restocking and rebarrelling, both functions outside the scope of this book. One major cost of restocking is the search for quality walnut blanks, which are becoming increasingly elusive and, consequently, expensive. With even a cheap double gun, where the stock blank is plain and therefore not very costly, the time and labour involved may still make restocking prohibitively expensive. With factory-produced singles, over-and-unders, self-loaders and pump-actions, factory replacement stocks can usually be obtained at realistic prices.

Rebarrelling is not something to be considered lightly unless it involves a pair of factory replacement barrels for an economy over-and-under. Even in that case, although proof laws between the continent and UK are supposed to be reciprocal, it may be difficult to get some cheaper barrels through UK proof when they have been fitted to the action. (The UK authorities seem to adopt a stricter interpretation of the proof laws.)

The rebarrelling of 'best' doubles currently costs around £6,000 to £8,000. If it is not carried out by the original maker, or under the auspices of the original maker, it can be less expensive, but not by a great margin. If a gun is rebarrelled by another maker it is permissible to fit the top rib with the original maker's name and address, but it should also bear the legend 'Rebarrelled by ... '. This may be tucked away on the keel plate but is still regarded as a valid means of identifying what is essentially a very major repair by someone other than the original maker.

It can make more economic sense to consider restoration rather than replacement. One method that has considerable merit is full-length sleeving of the barrel and chamber with a precision-made thin-walled tube. This is marketed as barrel lining and the present patent is held by Nigel Teague of Teague Precision Choke Manufacturing, although the origins of this type of repair can be traced to muzzle loaders rather than breech loaders. In the mid-1980s some muzzle-loading enthusiasts were reclaiming badly worn barrels by a similar method using hydraulic tubing, although the work was not advertised and not all examples were submitted to the Proof House; this is not an offence if it is work carried out by the individual on their own gun at their own risk. The big test, of course, is producing consistency of quality at a realistic price, and a method accepted by the Proof Houses that is also capable of withstanding the proof test. In this respect, Nigel Teague has succeeded where others have tried and failed. Barrel lining at the moment costs about a fifth of rebarrelling and the other significant advantage is that a gun with Damascus barrels retains its true appearance. At the moment the maximum degree of choke possible with this method is half choke, although work is in hand to produce tighter-bored chokes in the not-too-distant future.

The other method of barrel repair, which is more replacement than restoration, is sleeving; a means of reclaiming a gun with worn out or damaged barrels that was developed by the gunmaker E.C. (Chris) Ashthorpe in the early 1950s. It caused some controversy at the time, and this still rumbles on. Indeed, it has been described quite recently as 'effective if crude and fundamentally altering the integrity of the gun'. Much of this criticism is ill informed, as, done properly, it is a very effective and reasonably economic method of repair for about half the cost of barrel lining. Sleeved guns do, however, have a reduced value, particularly those that were stamped with 'sleeved' along the side of the barrel, at the insistence of the Proof House.

Sleeving requires the barrels to be cut off approximately 3¼in (8.5cm) in front of the breech and new barrel tubes to be spigoted in place back to the breech face. It is then rechambered and the ribs, keel rib and forend loop are replaced. As long as it is done with suitable finely made tubes (there used to be a supply from St Etienne that were very neat), the balance and handling characteristics of the originals can almost be duplicated. When it has been carried out by a really skilled craftsman, it is difficult to find the joint line, and with laser-weld-

ing facilities the joint line can be made completely invisible. Ironically, this method of repair is, in principle, the same as the modern mono-block system on which most over-and-unders and a few side-by-sides are constructed. Interestingly, such spigoted tubes will even work, albeit for experimental purposes, without being soldered in place. Chris Ashthorpe used to have a gun with the barrels just pushed in by hand. He would shoot it to demonstrate that the resistance on the external contact area of the spigoted tubes was greater than any tendency for the shot and wadding to push the barrels out of engagement. As for the integrity of a soldered assembly, the pressure required to push out a dummy sleeving tube from the breech end was, when tested, in excess of 26 tons per square inch. The proof load of most shotguns is around 3½ tons per square inch.

Common sense dictates that you are not going to sleeve a Purdey, Boss or Holland & Holland, but for a moderately valued gun it is a reasonable method of obtaining new barrels.

Of course, on the market originality and original finish are everything and sometimes a 'sleeper' does come to light – a gun that has lain unused, stored in good condition for perhaps fifty or seventy years or more. It may be dull in finish but otherwise as new, made when British gunmaking was at its peak; but it is always worth checking. As for that once-in-a-lifetime London gun described in the catalogue as 'retaining almost forty per cent of its original finish', that means that over sixty per cent of the finish is missing. After all, it depends whether you are buying or selling!

CHAPTER 4

Stripping and Servicing

Introduction

Stripping a gun for the very first time is an exciting and nerve-wracking experience, especially if it is a good example and someone else's property. Being mainly out of sight, the workings of a gun are often regarded as something of a mystery, even for someone who is reasonably familiar with the basic principles.

The initial enthusiasm to delve into this strange new world can be tempered by apprehension, imagining springs and unrecognizable parts leaping out to scatter on the bench, or, even worse, to fall on the floor and bounce away unseen into a dark corner. As you sit poised, tools to hand, uncomfortable memories may come back to haunt you – toys and old clock mechanisms taken apart and never quite put back together. And there was always the piece that was left over...

Fortunately, there is a logical order to both stripping and reassembling, and there are many similarities in the basic construction of sidelock, trigger-

The stripping area of the bench, with most of the tools ready to hand. Note the stripping block to the right rear of the rubber mat, the barrel-wall thickness gauge on the left and the bore comparator.

plate and boxlock guns. There are, of course, always detail differences, even with the same basic design, but nothing to cause any real mental confusion once the basic principles are understood and a little experience has been accumulated.

Before You Begin

The Working Area

Apart from those times when a complete strip-down is essential, guns often have to be part-stripped to effect repairs. It is both efficient and convenient to have a section of the available bench space dedicated for stripping and servicing. An area approximately 2ft (60cm) square is sufficient if a simple support is made to hold the barrels safe in a near-vertical position. Barrels lying across the bench are always vulnerable to accidental damage.

A raised edging in this working area is essential to prevent small parts rolling away, as they will do in spite of all the care taken to avoid this. A cut-out at one corner is necessary to allow brushing off the bench. Rubber matting fitted inside the raised edging is of benefit and a bonus if it is slightly ribbed – again to trap those parts that seem to have a life of their own.

It is very useful also to have at the back of this area a wooden block drilled to hold the most commonly used tools. The shorter screwdrivers or turnscrews can be kept blade uppermost for ease of selection. If they are far enough away, right at the back of the bench, customers will not be able to hurt themselves on the blades when they faint after being told the probable cost of repairs.

Safety: First and Foremost

Prior to doing anything at all with any gun, you must always check and double-check that it is unloaded. With magazine guns it is also essential to check that the magazine is empty and that there is not a cartridge jammed part-way in. It should become an ingrained habit so that, even when a gun is simply removed from the rack or store, the first move is to open the breech and check it is clear. At no time should the barrels be pointed towards another person.

Live cartridges should never be on the bench when working on a gun, especially loose and unboxed when dummy cartridges or snap caps are in use.

This may all sound obvious but strict adherence to those simple safety standards could prevent tragedy. Sloppy workshop practices and negligence may result in awful consequences, and even death.

The Sidelock

The sidelock may be a bar-lock or back-action lock, a hammer gun or hammerless (internal hammer) gun. Apart from detail differences such as ejector mechanisms, disc-set strikers, gas vents and assisted or easy-opening mechanisms, the basic order of stripping and reassembly is near enough the same.

1 Leaving the locks of a hammerless gun cocked, break the gun down into its major component parts: stock and action, barrels, forend. A hammer gun with a double cocking lock will need to be moved to the half-cock position prior to removing the locks. This is not, of course, necessary with a hammer gun with rebounding locks.

2 Remove the locks from the action. There may be only one pin (sometimes called a nail), but some guns have one long pin and two short screw pins behind the fences. Obviously, for those with hand-detachable locks, it is only a matter of taking out the one pin. Lock plates are often a tight fit but by screwing the long pin into the outside of its mating plate it can be used to pull the plate free. By inserting a small brass drift or punch through the sear arm cut out in the stock, the opposite lock can be gently tapped free.

3 Taking out the trigger guard screws is easiest done while supporting the bar of the action on a block. If the tang screws are unusually tight, the top of the stock (underside as you are working on it) can be supported with the bench horse opposite the screw to be freed. With the tang sprung out of the stock, the trigger guard is unscrewed anti-clockwise looking at it. On some guns this will require pulling the front trigger to clear the bow of the trigger guard as it revolves.

4 The next job is to remove the trigger-plate screw pin located at the front of the trigger plate and the long hand pin at the rear of the trigger plate. There are subtle variations on this arrangement. Sometimes what appears to be a tang screw is a hand pin that goes through the tang. Occasionally, the rear of the trigger plate is held by a screw like a tang screw. This is usually found where there is not a long hand pin fitted, the stock being secured by just the breech pin.

5 The long hand pin can (should) be quite tight and after unscrewing may be gently tapped out with a small piece of brass rod.

6 Holding the top lever to one side with the thumb, unscrew and remove the breech pin. The trigger plate can now be dropped out and the

stock gently dropped back then down slightly away from the action. The head of the stock where the wood is thin around the locks may have horns that fit into the action and care should be taken not to damage these. Never use force – if it seems necessary, something is wrong and a breakage of a delicate part of the stock is the most likely result.

7 Remove the lifters that cock the locks. If they are not marked, and reassembly will immediately follow cleaning, lay them out right and left on the bench with mating pins. Otherwise, mark them up accordingly with spirit-based felt pen or using masking tape and ballpoint pen. Using tape has the advantage that the pins can be held to their mating lifters.

8 Holding the action in one hand, push the top lever to its full extent and remove the top lever spring with the spring clamp. Sometimes the spring peg will lift straight out of the top strap and at other times it may need a gentle tweak with a small, fine-bladed screwdriver to loosen it. To remove the spring from the clamp the end of the arms can be gently compressed in a vice – a small hand vice is useful for this – and the clamp slipped off while covering with a rag to avoid the spring slipping out and leaping away. The vice is then slowly undone to release the tension on the spring arms.

9 Remove the safety button or thumb piece by pushing out the retaining pins or screw pin as appropriate. Where the safety spring is directly located under the safety button, this should be removed first.

10 Side-by-sides inevitably have a bar of some type to push the safety button back to the safety

Removing a top lever spring with the simple clamp.

position on the action of the top lever or cocking bolt before the gun opens. This safety bar may be located under the top strap, on the trigger plate or, with many imported guns, in a hole drilled in the stock. If it is of the type lying under the top strap, now is the time to remove it. If it is the sort that operates through a hole in the stock it may now be the time to search for it, as if one is unaware of this layout the rod does have a habit of dropping out unnoticed.

11 The top lever can now be removed by unscrewing the lever screw pin, which is usually very tight. Then the spindle has to be pushed out of engagement with the top lever. This is done by tapping or pushing it down by entering a flat-nosed punch down the threaded hole in the spindle. This is a much better option than tapping out the spindle from the top face – even with a brass punch, the end threads might be damaged. With the spindle withdrawn through the bottom of the action the top lever can be lifted out and the bolt withdrawn. Always check which way the bolt engages with the spindle as this is essential for reassembly.

12 Strikers are removed from the rear of the action either by first taking out locating screw pins or through the front of the standing breech via discs. Some guns have tubular screwed-in vents, which engage the discs, locking them in place, and these have to be removed first. The discs are removed with a disc-set striker tool fitted with the appropriate number and spacing of pins. It goes without saying all these parts must be kept in their matching pairs. Some gunsmiths will go as far as identifying each component with a unique mark, even down to bridle pins from locks.

13 The trigger plate can now be stripped. Where a trigger return spring is fitted, this should first be removed; note which direction it tensions the triggers. The spring is detached from the trigger blades by lifting and turning to release the two small prongs at the end of the spring arms.

14 The next move is to unscrew the trigger pivot pin (some cheap guns have a plain push-in pin) and withdraw the triggers from the bottom of the plate.

15 When stripping the locks it is a good idea to have two small boxes or trays so the parts are easily kept separate as right- and left-hand components. The first piece to remove is the mainspring, which should only ever be removed with a clamp, never using makeshift means such as pliers. With the lock cocked, only a little more

Parts should be laid out with components divided into right-hand and left-hand. If reassembly will not immediately follow cleaning, it is useful to mark the components.

pressure is needed to secure the mainspring in the clamp, allowing you to unhook it from the swivel and move sideways to detach the peg from the lock plate. Sometimes removal is aided, after clamping the mainspring, by disengaging the sear and intercepting sear, if fitted, to allow the hammer forward. When releasing the clamp tension on the hammer by unscrewing the clamp, cover with a cloth to keep it captive!

16 Remove the bridle by unscrewing the pins and gently ease it clear of the lockwork.

17 Remove the intercepting sear spring, which will be located by a pin and lug on the back of the spring, then lift off the interceptor.

18 Remove the hammer while compressing the sear spring with the sear, then remove the sear and sear spring.

19 Work on the barrels is fairly straightforward. Take out the extractor retaining pin, which, especially in the case of a non-ejector, may be located between the lumps or, as with ejector guns, a longer pin through the front lump. When this is removed, the extractor(s) can be removed.

20 Forends are secured to the forend iron in a number of ways but most commonly with a wood screw close to the knuckle, a pin engaging a fancy escutcheon on the front side of the catch assembly and sometimes one at the tip. With the Anson & Deeley pushrod tube, the pin may be hidden under the rod and spring.

Removing a mainspring from a sidelock with the mainspring vice.

21 When an A & D type is fitted unscrew the pushrod and spring after releasing the screwed locking pin.

22 If the forend tip is held on by an internal fixing, access to the screw pin will be through a hole in the pushrod tube. That is removed, followed by the wood screw in the forend iron stale (or steel). Lastly, the pin that is secured through to the escutcheon.

Removing ejector springs with my modified pliers, which give good control.

23 With a gentle tap against the pushrod tube the forend will now fall loose and the bolt that engages the forend loop can be removed from under the stale. (*NB: other methods of securing the forend, such as the old cross bolt or later Deeley catch, can be dealt with separate to the removal of the forend wood.*)

24 The ejector mechanism may be boxed and held by a screw pin, in which case it is usually best to release the spring tension prior to removal, the release of the kickers being done against a brass bar held in a file handle to ease the blow. If they are of the Southgate type it is easiest to leave them in the cocked position with springs compressed. To remove these is a bit of a fiddle so I devised a clamp using small pliers with the noses ground down to obtain access, and a bolt with wing nut through the handles to hold the spring in compression. As each spring is removed it is released under a cloth.

25 Tap out the kicker pivot pin and remove the kickers.

Once the gun is completely stripped, each component can be cleaned and examined for wear or damage. Small inaccessible holes can be cleaned out with a pheasant's wing feather. (Many other types of feather will do, but in the game season, pheasant's feathers are in plentiful supply. They are best kept in a sealed bag, otherwise they attract dust. Eventually they become brittle and fall apart in use, making more mess than effective cleaning, but a supply in late January should last until the start of the next shooting season.)

Most oil will have disappeared and old grease can be removed with a cloth dampened, if necessary, with methylated spirits. Any rust staining should be polished off either by hand using oil and 400-grade wet and dry paper followed by 1200-grade paper or, if necessary, on a rotary polisher by someone experienced and skilled in the mechanical polishing of small components.

Reassembly is simply the stripping or disassembly procedure in reverse. Remember to hold the lifters out of the way when putting the locks back in and make sure that the triggers are forward so they do not foul the sear arms. It is also useful to fit but not tighten the trigger-plate screw pin and long hand pin prior to fitting and tightening the breech pin. The trigger-plate screw pin and hand pin are tightened last, checking the safety works.

Each component deserves some protection against corrosion and lubricant for areas of contact. Moving parts, such as the lever spindle, the top lever where it enters the top strap, sear and bent engagement, ejector kickers, and so on, all require lasting lubrication. Oil does not last and to see an action liberally greased is not a pretty sight. Not only that, on a bitterly cold day a heavy layer of grease can slow up the lockwork even to the point of malfunction. The MOD-derived product PX24 is a good preservative, but may only be available in 25-litre drums. It is both a preservative and water dispersant and dries leaving a thin and invisible chemical coating; this does not immediately inspire confidence, but the product is superb. To test its efficiency as a preservative I placed steel blocks fresh from the grinder on the outside window ledge of the workshop protected by various methods, including PX24, oil and grease. It was winter and for three weeks there was alternately rain and sleet with some snow and a liberal helping of frost. The invisible barrier of the PX24 proved to be one of the most durable protective coatings.

However, PX24 does not lubricate. A fine graphite-based lubricant that still works when dry is a very suitable addition to the PX24 for gun locks when applied sparingly to the moving or contacting parts. Also it stays where it is put and does not leach into the stock wood. An economic alternative to PX24 is silicon lubricant. If this is applied to an

old duster and left for a few days, a barely damp duster is the result. The silicon-impregnated duster can be used to wipe over all the individual parts during assembly prior to adding lubricant.

Anson & Deeley Boxlock

Much of what applies to the sidelock also holds true for the boxlock, particularly the manner in which the stock is removed from the action and stripping the forend assembly and extractors from the barrel. The big difference, of course, is the lockwork, which is retained within the body of the action even when the stock is removed. Also most boxlocks have a bottom plate that is essential to access the locks. The differences in stripping are as follows:

1 When removing the trigger guard the locks of a boxlock should be in the fired state, and this is, anyway, essential for later disassembly work. If they are not, this can be effected by pulling the triggers with the breech face against a piece of hardwood or the edge of a wooden bench to cushion the hammer blow. If the breech has side-clips a piece of hardwood can, instead, be held against the breech face.

2 Remove the trigger guard, which again may require pulling the front trigger out of the way of the bow at least for the first couple of turns.

3 Remove the screw pin and bottom plate (something a sidelock does not have this). If the bottom plate is tight it can be gently tapped free by inserting a piece of brass or dowel between the action slots where the lumps engage.

4 The removal of the stock by unscrewing the trigger-plate screw pin, hand pin and breech pin is the same as on a sidelock, except that the boxlock stock is a little sturdier and does not normally have projecting horns.

5 If, for some reason, the action has been left cocked, it can now be let down into the uncocked or fired position. This is done by pressing down on the end of the hammer with a brass bar fitted into a file handle (a most useful item), releasing the sear, and releasing the hammer to the uncocked position under hand pressure.

6 Now the real differences between the sidelock and boxlock become clear and it is possible to start to understand the ingenuity of Messrs Anson & Deeley and their tremendous contribution to the world of the side-by-side shotgun.

7 First, remove the sear springs, which in an Eng-

lish-made boxlock are flat steel strips retained by a screw pin. On some imported guns they may be coil springs and plungers held within the sear body and pressing against the rear of the action. The former can be removed prior to taking out the sears; the latter cannot.

8 When the sear springs have been removed, place the action on the stripping block so that the plain sear pin lines up with a hole in the block and tap it completely out of the action with a flat- or concave-nose punch. The sears can now be lifted free as the punch is withdrawn.

9 Next, remove the cocking dogs – or limbs that pivot on pins in the front of the action bar behind the line of the cross pin. Sometimes these pins may be locked in place either with a small short screw pin that engages into the head of the pivot pin, or by small headless screw pins inside the bottom of the action. Remove any locking pins and withdraw the cocking dogs individually.

10 Next comes the tricky bit, one of the peculiarities of the boxlock: removing the hammers, which, of course, still have spring tension against them. Unlike the sears this is a job best done one side at a time and with the action secured in a vice using soft jaws. With an undersized punch, gently tap out the plain pivot pin a little over halfway and then the hammer will be felt to move on to the punch as the pin disengages. The A & D mainspring compressor, in this case a piece of wood shaped in a curve that fits the action slot and is equipped with a handle, is used to push against the hammer while imparting a twisting motion to draw the hammer away from the back of the action. As the spring tension is taken up on the compressor, the punch can be withdrawn and the hammer and mainspring gently released. Then it is a matter of reversing the action in the vice and carrying out the same procedure from the other side. The punch should never be removed with the hammer still under tension from the mainspring; they are quite powerful and letting the whole lot fly free could result in an unpleasant injury.

11 The rest of the action, top lever, safety, and so on, can be stripped out in just the same manner as the sidelock. There are no great surprises.

Reassembly from a selection of cleaned and carefully laid out (right- and left-handed) parts is more or less a reversal of the stripping. It is useful first to refit the bolt, spindle and top lever to the action as it is not always possible to fit the spindle with the sears and springs in place. The other problem is refitting the hammers against the tension of the mainsprings. This is another job to hold in

Replacing the hammer and mainspring from a boxlock. Sometimes this is a job that seems to need three hands.

between the vice soft jaws. With the spring engaged in the hammer, the two parts are entered into the action slot and pushed into place with the spring compressor. When, as with many English guns, the striker is an integral part of the hammer, this can be used to some small advantage. As the hammer slides around the back of the action the striker will engage into the striker hole, which provides some positive location. Then it is a matter of manoeuvring the hammer until the pivot hole lines up with the hole in the action. It is very difficult to get exact alignment but if a taper-nosed punch (with a short sharp taper), with its major diameter the same as the diameter of the pivot pin, is inserted it will align the hammer and hold it in place until the proper pin can be tapped into place from the other side. You can then drive the pin on through to leave the opposite action slot clear for refitting the other hammer and mainspring.

With many imported boxlocks the job is much simpler as the action slots are parallel through to the knuckle of the action, unlike the stepped slots of which so many British makers seem so fond. With these, the hammers can be fitted in place and the mainsprings dropped into the action with room for the vee of the spring to project back through the knuckle. By pressing down the end of the spring with a piece of brass bar fitted in a file handle, and cut with a cross slot to engage the spring, it can be pushed into place under the hammer. This is a much simpler and safer way of fitting a mainspring in a boxlock and it is hard to understand why British makers rarely adopt this simple idea.

Once the hammers and mainsprings are in place, the tricky bit has been completed. The only other fiddly bits worthy of mention are sears fitted with helical springs and plungers. These have to be compressed against the back of the action while the sear is aligned with its pivot hole. This, again, is where the taper-nosed punch is useful to obtain the alignment, so the plain pin can be tapped through without catching on the edge of the sear pivot hole and jamming up.

After reassembly of lock, stock and barrel and prior to putting the gun back together the action should be cocked. This is done before fitting the bottom plate, by pressing against the cocking dog where it contacts the hammer and pushing until the sear engages. Yet again, that deceptively simple multi-purpose tool, the brass bar fitted in the file handle, comes into play. (How is it that something so useful does not have a name – how about 'the nudger'?)

It is most important to do this with a gun on which the strikers are part of the hammer as they will, when at rest, protrude through the breech face. Then, when attaching the barrels, the extractor will be partly pushed out and can neatly chop the ends off the strikers if the hammers are not cocked.

The Trigger-Plate Action

Once you are familiar with the sidelock and boxlock, the trigger-plate action really holds no great surprises, except for the fact that it does not always have a breech pin. The lockwork is attached to the trigger plate, which is secured to the action by one or more screw pins and through the stock by a hand pin. With the trigger plate freed, usually with the trigger guard still attached, it is dropped away from the stock and action. It is then only a matter of withdrawing the stock from the top strap.

The action is amazingly simple, usually quite slim and this, apart from the lack of a breech pin, is a good visual first indication that it is not a boxlock.

One advantage of incorporating the locks on to the trigger plate is that the actions can be well rounded, in the manner of the much-acclaimed Dickson Round Action and its derivatives, which are noted for their fast, fine handling qualities. Mainsprings can be a bit of a fiddle to remove as it is sometimes difficult to get the clamp under the bottom arm of the spring where it runs close to the lock plate. A modern-made economically priced reproduction spring clamp can be slimmed down and kept especially for the occasional trigger-plate lock job.

The hammers and sears are held in place either with shouldered screwed pins or plain pins, which, once detached from the mainspring, can be easily removed.

The trigger-plate action is an intriguing design that, while not as simple as the boxlock, is actually easier to work on and lacks the multiplicity of parts that adorn some sidelocks. Some have the ejector springs carried in the bar of the action, which seems a good idea and takes some of the load away from the forend that occurs with more conventional systems. The one weakness of this design is the amount of wood that has to be removed to accommodate the locks, which certainly leaves it weaker than the boxlock or even a sidelock stock. To a certain extent a long top strap and trigger-plate tang help rectify this small deficiency and while, like so many things, this may be regarded as imperfect in gun design, it is none the less a beautiful kind of imperfection that has many more virtues than vices.

The Boxlock Over-and-Under

The 'boxlock over-and-under' bears no relationship to the Anson & Deeley boxlock; it is simply a convenient way of describing a particular design, typified by the coil mainspring inertia block system located between the fixed arms of the top and bottom tang of the action or frame. In some ways it is more akin to a trigger-plate action but, except for those like the Beretta, it does not properly qualify to be described as such, so a 'boxlock' it is.

These boxlocks do not have quite the conformity of design that might be expected with more traditional guns, so the detail differences are more noticeable. As a result, any description of stripping the boxlock over-and-under as a type must, of necessity, be a little less detailed to allow for variations in design. Nothing of startling complexity is likely to be encountered — after all, most of these guns are produced in a manner that is intended to be convenient for manufacture — although some can be quite odd. There is at least one gun where the top lever spindle also serves the function of retaining the firing pins while most have plain pins to secure them in place. The majority of boxlock over-and-unders have coil top lever springs, but one make uses a rather short vee spring located at the bottom of the top lever spindle, and some are equipped with short vee mainsprings that often seem barely up to the job.

A Browning 325, one of the most popular boxlock over-and-under actions.

The actions and lockwork of the better makes are sturdily made from investment castings and the internals are given a protective coating. Lesser guns vary from adequate to awful, some with internals that appear to have been the leftover parts from a dramatic failure at the Proof House. Some of these economy guns were probably chasing a once-expanding market for over-and-unders and the external appearance and, sometimes, the original retail price do not always give much of a clue to the horrors that lie within.

1 Once the gun has been broken down into its three main component parts — stock/action, barrel and forend assembly—work can begin in earnest. With the B25 Browning, where the forend only slides forwards to remove the barrels, it can be completely removed by accessing the screw pin located under the forend catch with an angled-shank screwdriver.

2 The method of removing a butt plate to access the stock bolt is obvious, but a lot of over-and-unders will be fitted with butt pads, either as original issue or retro-fitted. Most will be retained with Posidriv, or what used to be called Phillips, screws of the cross-head type. The better pads have cylindrical inserts, which can be prised out so the head of the screw is visible. Others, if particularly neatly fitted, will have no obvious means of fixing, although with careful examination a small slit can be located. Prior to pushing a round-shank screwdriver through the butt pad it should be lubricated to avoid tearing the rubber. Washing-up liquid is a good medium as it can be easily cleaned off, leaving no trace. It is worth noting that a few competition guns are supplied with tooling that enters to fit the stock bolt through a hole in the butt pad.

3 With the butt pad removed the stock bolt can be undone to release the stock. Some bolts have a plain cheese head (with single slot), some are slotted hexagonal heads and others are hexagon socket heads or manufacturer's specials. Most are difficult to see and a variety of tools have to be available. Whenever a hexagonal head is fitted, whether slotted or not, it is best to use a small socket on an extension as it gives a more positive feel than a long screwdriver. Always check prior to undoing the stock bolt that there are no screws retaining the trigger-guard tang to the stock. If there are, these should be removed first.

4 With the stock removed the action is exposed and if still cocked the hammers should be let down into the uncocked position. Where the

Partially stripped boxlock over-and-under, which has fewer parts than a sidelock.

hammers have a hole through them they can be held with a small piece of bar and let down gently after tripping the sears. To remove the mainsprings it is a matter of using round-nosed pliers to push back on the spring guide to compress the spring and then releasing the end of the guide sideways from the back of the hammer.

5 The hammers can now be removed by pushing out the pivot pin. Where ejector trip rods that engage a lug on the hammer are fitted, these are removed first.

6 It is now useful, with the hammers out of the way, to remove the firing pins. Most have return springs and are secured by plain cross pins. Some top firing pins do not have return springs and these are identified by having a longer body. It therefore follows that, with some makes, even those where both firing pins are sprung, the top and bottom pins are not necessarily identical.

7 By first removing the selector (or safety) spring, the selector safety, as the safety is often called, can be detached, leaving the inertia block free to flop about.

8 With the trigger guard removed, and one of the plain pins holding the trigger to the inertia block connector tapped out, the trigger may be withdrawn through the bottom tang and the inertia block and connector lifted out.

9 Cocking rods that lie in the bottom of the action are sprung and some have a cam between them and the hammers. With this layout the cam pivot pin is also the release for the return spring. These cocking rods are withdrawn from the front of the action, which sometimes requires the release of the bolt to clear their removal.

10 The top lever is invariably manufactured as part of the lever and spindle assembly with cut-outs in the spindle to clear the firing pins. Once the return spring has been located and removed – this may even be a spring bearing against the locking bolt – the spindle and lever can be withdrawn after tapping out the plain locating pin.

11 Ejector springs on over-and-unders are either incorporated behind the extractors or operating kickers in the forend. Those in the forend can be released in the same manner as the mainsprings with the kickers – or hammers as they are termed on many over-and-unders – first

61

released to the fired position to reduce the spring tension. Ejector springs incorporated behind the extractors are released by removing the extractors. With some makes this is achieved by pushing down a little way and twisting the extractor leg to one side, whereupon it disengages from the dovetail slot. Others, where a sliding bar engages up into the leg, may be released by pushing down through a small access hole in the sliding bar to depress a plunger and spring, enabling the bar to be disengaged from the extractor leg. It is important with any of these types to remember that the extractor is spring-loaded so caution should always be exercised when releasing the extractors. It is beneficial to release them by holding the barrels with the extractors partly compressed against the front of the bench so that the spring pressure can be released slowly.

12 The travel of non-spring extractors is limited either by a screw pin behind the extractor extension or a projecting steel roll pin in the top far end of the extension.

Once all the pieces – some of which will be fairly unfamiliar to some gunsmiths – are laid out on the bench, it is just a case of cleaning, lubricating and remembering where all the bits came from. If they are laid out in logical order at the time of dismantling, this should not be at all difficult, as none of these designs is really complicated.

The Semi-Auto and Pump Action

Without doubt, the majority of traditional gunsmiths view the arrival for repair of either a semi-auto or pump-action shotgun with a lack of enthusiasm, while many approach such a job with loathing.

Dealing with these guns requires a different state of mind; regarding them as a sort of machine tool designed to feed, fire, eject and re-feed cartridges is a good start. The ability to think three-dimensionally is useful as is the need to remain calm when considering all those springs, pins, pressings, and other unfamiliar parts. For many gunsmiths, the unfamiliarity is the crux of the problem – these guns are different, and do not have beautifully hand-finished parts that we can admire.

As with the boxlock over-and-under, the principles of operation are similar, in spite of the fact that semi-autos are either gas- or recoil-operated, and pump-actions are manually operated. Some are very similar. For example, the well-known Remington Model 1100 semi-auto has a pump-action stablemate, the Wingmaster, which is externally almost identical and also shares most of the same components.

1 The first move in any strip-down is to remove the barrel and with semi-autos the forend. This requires the mechanism to be locked back in the 'fired and empty' position to release the breech block, sometimes termed the breech bolt, from its engagement with the barrel extension. Then the barrel can be released forwards by unscrewing the magazine cap. With recoil-operating semi-auto guns it is necessary to apply some hand pressure on the barrel, pushing it towards the receiver to ease the pressure on the recoil spring to be able to undo the magazine cap. With the magazine cap taken off, the barrel and spring can then be released under full control. Taking off the magazine cap may also release the magazine spring but there should be a separate cap fitted internally into the magazine tube.

2 The next job is to release the mechanism forwards by pressing the release catch and letting the breech block forwards by hand, using the cocking piece or operating handle to control this. The trigger mechanism can then be taken out of the receiver, which, while carried in a block, is simply a type of trigger plate. Some are retained by a lug at the rear and a screw pin at the front, others by two push-in plain pins and similar variations. This is a job best carried out with the gun upside down so the mechanism is lifted out. To avoid dropping any spring latches that lie along the inside of the receiver, hold it in place by the trigger plate and pins.

3 As the trigger plate is lifted out it is worth noting which way up the spring latches are fitted. Those types that are unsupported without the trigger plate in place should be marked up in the direction in which they fit.

4 The method of removing the breech block is one of the differences between the semi-auto and the pump-action. Almost all semi-autos have a cocking lever or operating handle, which, with the breech block rested forward in the action, can be pulled out sideways. The breech block can then be drawn forwards through the action, along with the action bar and sleeve, which is released from the breech block once it is clear of the magazine.

5 With most pump-actions the action bar is fixed to the sleeve making an action slide assembly, which is retained inside the forend by a tube nut

in the form of a screwed ring. With the tube nut unscrewed it is possible to slide off the forend and expose the magazine and action slide.

6 Many pump-actions have a stop ring that limits the forward travel of the action bar assembly, requiring the magazine to be unscrewed to release the action bar and breech block. With others, the bottom part of the breech block can be detached from inside the receiver by aligning it with slots cut in the side and thereby releasing both the action bar and breech block proper. Sometimes the elevator assembly that carries the cartridge up from the magazine has to be detached to achieve the removal of part of the whole of the breech block.

7 The one extra in the semi-auto is the action spring, which is the return spring for the breech-block assembly and lies in a tube down inside the stock. This can be accessed by removing the butt stock, the link between the spring and breech block being removed from inside the action by raising the tail and twisting slightly.

8 The breech block of both semi-autos and pump actions can be stripped separately on the bench. This contains the firing pin, firing pin return spring and locking block, as well as sundry plain pins to hold the assembly together.

9 The trigger-plate actions of these guns, to which the cartridge carrier is also attached, are, if in good condition, most suitable candidates for the ultrasonic bath followed by blowing off with an air line using a suitable safety nozzle. Rarely do they need a complete strip-down – after all, they are not that delicate – but they do need to be lubricated sparingly, then reassembed and tested.

There are benefits to owning a semi-auto or pump-action if it is of a reasonably modern vin-

The main parts of a semi-auto action: receiver, bolt and trigger plate from a Breda, one of the lightweight semi-autos.

tage, because instructions on stripping and exploded diagrams are available. So even if you get a little lost on reassembly, it is not a complete disaster!

Some Problems With Stripping

The most common problem occurs with traditional hand-finished guns where the pins are tight and the slots in the heads very narrow and sometimes rather shallow. It is difficult to get an adequate purchase on some of these, even with a correctly fitting screwdriver or turnscrew. One way is to support the job in the vice or on the bench horse and, with a screwdriver of adequate length, literally get your shoulder behind the hand holding the handle, and apply some weight. This can be increased by standing on one leg; the body load that would be taken by the raised leg is transferred to the job, and the hand or hands are left free to apply the turning moment.

If the problem is getting enough turning moment, a brace holding a short bit can be used or a reasonably sized tap wrench fitted to a square-shank screwdriver.

If a screw pin is exceptionally tight, sometimes a blow using a piece of brass bar and hammer will provide enough shock to release the threads. One of the tightest is often the long hand pin, partly due to the ingress of water and subsequent fine rusting down the thread where it mates with the top strap or tang. One means of getting such a pin out is to support the top strap or tang against one vice jaw while the other jaw clamps up against a short screwdriver bit with a square or hexagonal drive end. With pressure applied by the vice and a spanner to the bit drive end, the thread can be tweaked just enough to start it coming undone.

Where it becomes impossible to remove a pin, either perhaps because it is really rusted in place (the hand pins of the little Belgian folding .410s were particularly susceptible to rusting in place), or because the slot in the head of the pin has been previously and extensively damaged, the remaining option is to drill it out. With a hand pin the threaded end is visible and a small hole drilled down the centre of the pin for about the length of thread engagement, and then gradually drilled out to approximate the thread root diameter, will release it. After this, it is a matter of making a new pin, which is only complicated by the need to identify the original thread and choose the right material. Pins of all types are tough and mild steel will not do, unless it is case hardened and that, if overdone, will result in brittle threads. One colleague of mine swore that the best material came from the old spring tines from a hay-turning machine. I tried this once, with little success, and since then I have stuck with EN16, which is sufficiently tough but reasonable to work with. It also has a high resistance to any shock loading – it is the same steel used for the spindles in vintage motorcycle girder forks, which take tremendous punishment.

CHAPTER 5

General Barrel Work

Introduction

A good pair of side-by-side barrels, neatly struck down with square edges to the ribs, nicely squared-off muzzles and hand-polished breeches is surely a thing of beauty. When the barrels first come into the workshop, they may be knocked and dented, rusty with blacking or browning worn and rubbed and perhaps the ribs loose – a picture of abject ugliness. Comparing the former with the latter is rather like comparing the latest showroom offering with a shabby older-model car.

It is a mystery why so many gun owners neglect their barrels, the most important part of their prized possession, which is also one of the components most vulnerable to damage. Most are very keen on seeing a pit-free, shiny bore, but display a rather laissez-faire attitude to other important, or even potentially dangerous barrel damage. The gun owner who wants the gunsmith to remove a couple of small pits by boring out 99.9 per cent of the corresponding barrel wall often expresses surprise at the number of dents that first need removing. Modern barrel steels are very strong and those with hard chrome bores are virtually corrosion-free, but they are not in the majority and there will always be some form of barrel work to be carried out.

Dent Raising

'Just purchase one of those super modern hydraulic dent raisers and pop out the dent.' At least one advertisement used to suggest that raising dents was that easy. It is not, of course. When a barrel wall is dented it both deforms and stretches the metal and somehow most of this has to be persuaded back into place by applying a form of swaging.

First, it is necessary to locate the dent or dents, and mark with a circle on the outside with chalk (hard engineer's chalk is best) if the barrels are

blacked, or spirit-based felt marker if they are in the white. If the barrels require refinishing, an easy way to identify the exact location of a dent is to go over the area with 400-grade wet and dry paper on a flat block. This will remove the finish where the barrel wall runs true, while leaving it untouched and very visible in the dent. Initial alignment of the shoe of the dent raiser is done against the outside of the barrel. Set the stop so the depth of engagement into the barrel is correct and use the locking screw to align with the shoe that is going to contact the dent. Slacken off the adjuster so the shoes are smaller than the bore and make sure they are clean and lightly oiled, then insert full length to the adjuster. Line up by eye the adjuster locking screw and the dent.

The adjuster is turned until it contacts the dent and a little extra pressure is applied – the exact amount of pressure is really a matter of 'feel' and experience. It is wise to practise on scrap barrels, as this is a tricky process and without great care barrels can be ruined. Damascus barrels are relatively soft and some steel barrels feel little different, while later barrels of higher tensile steel or heavier wall thickness require more pressure.

The metal then has to be re-formed and for this a special hammer is used. One much-used type has a rounded nose with the appearance of a small pick.

Dents in a Damascus barrel, particularly obvious after cleaning off old rusting.

Modified cross pein hammer for dent raising. The modifications of the cross pein to a 'pick' form are obvious. Not so obvious is the reshaped semi-hemispherical form of the larger end.

Find an old pair of barrels where the dents have been raised but no refinishing has been done and an unmistakeable patch of small peening marks can be found where there was once a dent. I use a converted panel-pin hammer with the nose well rounded and polished, and the cross pein converted to the 'pick' form.

Tapping around the outside edge of the dent will give the effect of loosening the dent raiser shoe as the barrel wall starts to change shape. Retighten the dent raiser and go around again. Gradually it is possible to work around the dent, moving inwards, until the whole area is covered. When the shoe sits flat against the inside of the barrel – this can be checked with smoke blacking – the outside can be peened in line with the axis of the bore. It cannot be stressed enough how important it is not to apply too much pressure and bulge the barrel.

Some barrels, particularly those of later manufacture and those that are damaged only with a simple small dent, can often be re-formed back into place leaving hardly a trace of the damage. On softer barrels (a relative term, of course), and where a dent may have metal bulged to the side, it is not possible to re-form it all to leave a perfect bore. To do the job properly the bore should be lapped or honed to leave it true, and the outside of the barrel, especially if multiple dents have been attended to, should be struck up and refinished.

Bulges can usually be swaged into place more easily than lifting a dent. The principle of supporting the area with the dent raiser is similar but the larger end of the peening hammer formed in a curved shape is best used to dress the bulge down, this time gradually working from the middle outwards.

Rib Repairs

Ventilated Ribs
Raised ventilated ribs on over-and-unders are easily deformed and fortunately just as easily repaired. Small tapered brass wedges inserted in the gap under the rib are used to lift the deformed or bent section back into place. If there is sufficient gap a pair of wedges is ideal, giving a straight push, but in practice one wedge works well enough.

Minor Rib Repairs
When main ribs become loose, even over a small area, there is only one answer and that is to strip them off and relay them as a complete job. The exception is a forend loop pulled loose and the short rib (keel rib) lying between the forend loop and the barrel flats. Sometimes the forend loop and rib are one piece, and in this form they are less likely to loosen as the contact area holding the actual forend loop in place is much improved over an assembly of separate parts.

With the forend loop, as with any other soldered or what was once called sweated joint, the fit of the component parts is of utmost importance. Solder, an amazingly useful amalgam of lead and tin, has little inherent strength and simply using it to fill gaps is a waste of time; the better the mating of the parts and the thinner the solder line, the stronger the joint. When a forend loop is found to be badly corroded and deeply pitted, thereby reducing the

Far too much solder – even after some has been removed. An amateur repair using a large soldering iron with insufficient heat input.

effective contact area, it may be necessary to make a new one. If the corrosion appears extensive, and even though the rest of the ribs may be firm, it could well be best to strip them off and check the barrel section hidden beneath the ribs.

Assuming the forend loop is in good condition it is necessary to clean off any excess solder and then smoke black the tongue of the loop to check contact with the barrels. If this proves unsatisfactory only a small amount of reshaping with a file is possible before the forend loop pushes down too far between the barrels. Cleaning off all traces of solder, building up with weld and refitting is one answer, which avoids making a new one.

With luck, the forend loop is in good condition and, when smoke blacked, shows good marking at all point of contact, such as between the barrels and either side between the bottom rib and keel rib. I think it is always best to re-tin at least the tongue of the loop and the area between the barrels, unless there is still a layer of solder in this area. If there is, it still needs to be reduced to the minimum and the best tool to use is a small white metal scraper. When a part has to be re-tinned the most efficient means is the use of solder paste, sometimes called solder paint. It can be applied to the part, then heated so that turns into solder; if this is wiped over with wire wool dipped in resin flux, a beautiful clean tinned area is produced. It is, however, corrosive! Any parts treated with solder paste should be thoroughly washed off with warm soapy water to eliminate any possibility of ongoing corrosion after assembly. It is certainly most undesirable simply to paint and assemble the parts with solder paste and then heat them until a joint is formed, because chemical corrosion between the barrels can take place.

Once the parts are tinned and clean, treat them to an application of non-corrosive resin flux prior to assembly. A small amount of powdered resin inside the immediate area is not a bad idea either, as it melts and will form a protective coating on the previously heated parts. Heated steel, if unprotected, is notoriously susceptible to after rusting.

Subsequently, it is a matter of wiring the existing ribs to prevent them becoming loose, as well as wiring in place the forend loop and inserting heat sinks into the barrel to reduce the spread of heat to the rest of the ribs. With this preparation done, a small nozzle in the propane torch is used to heat the loop. One useful trick here is to place two small pieces of fluxed electrician's solder either side of the forend loop so that, as it heats up to a temperature that will form a soldered joint, this solder will melt and run along the joint as a visible sign that working temperature is being reached. Another small trick: when you are satisfied that the joint is completely at temperature, apply a little more fluxed solder around the edges while pressing the loop firmly in place with a piece of hardwood. This is necessary for two reasons: first, as the wire holding the forend loop into place is heated it will soften and expand, reducing the pressure it was exerting holding the loop in place; second, as the tinning melts on the mating parts and more solder is introduced, the tongue of the forend loop will try and float on a build-up of molten solder. By applying extra pressure to the forend loop the small surplus of solder is pushed out between the mating parts,

Everyone has their own ideas, but this seems to work well and is easily made from old lengths of rifle barrel. In use, the flat section is turned adjacent to the rib being heated so it does not draw the heat away; the round part is against those areas to be kept cool. The handles are marked in line with the flat.

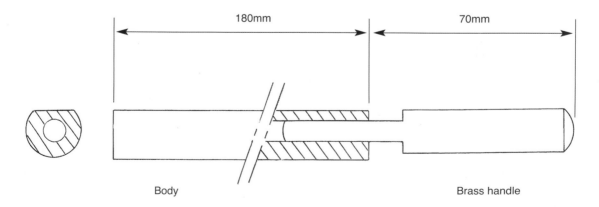

180mm   70mm

Body   Brass handle

to ensure minimum solder in the joint and maximum strength.

If too much solder is applied it can squeeze out to form a blob of loose solder in between the ribs, which later, as the barrels are raised and lowered, rattles up and down in a disconcerting manner. Drilling a hole in the bottom rib and squeezing in grease to trap the offending solder is not the answer. If you are lucky, reheating the loop (with everything still wired in place) can melt the loose solder back on to the tongue and rid you of the 'rattler'. Failing that, it means taking out the forend loop, clearing the surplus solder and learning from experience!

The Muzzle

Muzzles are a little vulnerable to damage, especially if they are already fairly thin. To trim and true up muzzles a barrel-facing tool is used; the removable pilot fits into the muzzle and a rotary cutter trims the barrel end. These tools are available from the USA or can be made from a piece of chrome molybdenum steel such as a piece of the breech end of a full-bore rifle barrel (not black-powder as they are usually softer material), suitably hardened by heating to cherry red and quenching in oil.

Trimming a double leaves some overlap into the other barrel so it is a case of cutting a little at a time on each barrel then finishing with wet and dry

Simple muzzle-trimming tool made from old chromium molybdenum rifle barrel. The teeth are hand cut, then hardened by heating to dull red and quenching in oil. A different pilot is needed for varying bore sizes.

held against a flat bar or ground-down file as a suitable backing. With a single-barrel gun no such difficulty arises and, with care, plus a little lubricant such as tallow, a good finish can be produced with the cutter.

Barrel Choke

There is a fair amount of truth in the old saying that choke lengthens your reach but lightens your bag. Ideally, the shooter should use the minimum amount of choke compatible with the distance at which the shot is to be taken and the size of the quarry. Much of the time with live game shooting it is inevitably a compromise, as circumstances are rarely as planned. With clay pigeon shooting in a controlled environment it is quite a different matter and the majority of competition shooters now use interchangeable screw-in chokes, suitable for any discipline.

Many game shooters, particularly those devotees of the traditional side-by-side, are not at all enthusiastic about having a gun fitted with screw-in chokes – even supposing the barrel walls are thick enough to accommodate them. In fact, I think is it fair to say that most would regard such a conversion as about as appropriate as harnessing a racehorse to a plough. In simple terms, choke is the restriction at the muzzle that concentrates the shot pattern. The effect of this as viewed by the shooter is, as the choke is increased, to be able to hit the quarry hard at a greater distance or enable the average shot to miss easier at close range. Conversely, few game guns are bored true cylinder as the shot pattern spreads very quickly and, to ensure a clean kill, is little use except for close shots such as those made in thick woodland. In reality, to improve even the pattern of an open-bored gun, it is useful to retain a few thousandths of an inch of choke.

The standard test range for checking shot patterns is 40 yards/m, which to most shooters seems a long way. Occasionally the gunsmith will be requested to bore out the chokes of a gun to produce a pattern with given cartridges of a certain shot size.

Most of us who have attended game shoots have witnessed the 'expert' who pokes a finger down each of the bores of a double gun and declares in loud and satisfied tones the amount of choke. This is, of course, nonsense. Choke is the amount of restriction relative to the main bore size. Nominally choke ranges from cylinder bores (no choke) to full choke (see table below) and the main bore size is not always what might be expected. Take the ubiquitous twelve-bore. Many older guns were very tightly bored so on occasion you may find a gun chambered twelve-bore with thirteen-bore barrels, .710in. Compare this with a 'true' twelve-bore that gauges at, say, 0.730in.

Chamber	Barrel bore	Measured full choke size	Choke
12-bore	13 (0.710in)	0.670in	0.040in
12-bore	12 (0.730in)	0.690in	0.040in

Therefore the difference is 0.020in, even though both guns are twelve-bore and nominally full choke.

From this simple example it is easy to see that choke, to whatever degree, is never going to be a predetermined range of sizes. There are also other complications. A gun bored at the lower end of a bore size with a given amount of choke will not pattern quite the same as a similar gun with the same restriction of choke and within the same size range but bored more openly. Take this to extremes and twelve-bores may be found with barrel sizes ranging from .710in to .740in. Bores may also be measured and found to have some taper apparent. Spill boring barrels often produce taper of perhaps 0.002/0.004in, which is generally regarded as beneficial but also contributes towards the resulting pattern. Choice of cartridge, shot load and type of wad – fibre or plastic cup – all contribute to the end result with a particular gun, and this is where the use of a pattern plate is important. As an example, do not be surprised if, when patterning a 30in barrelled twelve-bore on the lower end of nominal bore size, with some taper and a comparative measurement of quarter choke, to find it produces a tighter pattern than expected, maybe even nearer half choke.

Choke is measured in the UK in thousandths of an inch or what was once called 'points' of choke. It is easy to fall into the trap of assuming that if 0.040in – 'forty thou' in common parlance – is full choke, then 0.020in is half choke. This may or may not be so, as one gun might produce full choke patterns with less than 0.040in, while another might need more than 0.020in restriction to throw half choke patterns.

Still, we have to start somewhere and the following table of choke is a useful guide, albeit if only a nominal guide towards the end result:

Full choke	0.040 inch
Three-quarter choke	0.030 inch
Half choke	0.020 inch
One quarter choke	0.010 inch
Tight improved cylinder	0.007 inch
Open improved cylinder	0.003 inch
Cylinder	as bore size (rare)

Some Advice on Chokes

Many mass-produced guns leave the factory with an abundance of choke. This at least gives the owner the chance at a later date to tailor a gun to his needs and many will turn to a gunsmith for advice. For driven game shooting most users are well served with a combination of quarter choke and improved cylinder. This is also useful for pigeon decoying and duck shooting over flight ponds. If the user feels that for duck more striking power is needed, it is better to recommend going up a shot size, where the pellets carry more energy; in the UK, lead may no longer be used for shooting wildfowl. For duck and geese on the foreshore it is a different matter as shots are often taken at longer range with heavier-loaded cartridges using larger shot sizes. In these circumstances, guns are often bored full and half choke, or even full and full.

In going up in bore sizes there are other factors to consider. A big-bore gun – eight-bore or larger – throws more shot and a larger open-bored gun in practical terms still has more effect than a tightly bored twelve-bore with magnum cartridges. My comparatively light single-barrel muzzle-loading four-bore with a trace of choke is loaded with 3¾oz (100g) of shot. For any given distance over which a twelve-bore might be used, the four-bore obviously has a distinct advantage as it is pushing out three times the amount of shot. Even allowing for the greater spread of the nominally improved cylinder barrel, the density of shot for a given distance is still greater than that of a twelve-bore with a lot of choke.

The only thing that is certain is that, whatever a gunsmith can do to improve a shooter's chances of success, it will not necessarily improve his aim. There are, after all, really superb shots who can use a gun carrying a lot of choke and consistently clip driven pheasants in the head, or even effortlessly and neatly take running squirrels from the branches. The rest of us need a little help, however, and that is achieved by keeping the choke fairly open and the pattern reasonably large. Even so, on those days – everyone has them – when a Nock seven-barrel volley gun loaded with shot would prove insufficient, it is not the gun or the choke that is at fault.

Choke Evaluation by Patterning

Equipment A pattern plate is a substantial device of about 8ft (2440mm) square boarding faced with, as a minimum, sixteen swg steel sheet. Most of us do not have this luxury although a local shooting school might have one and be prepared to come to a mutually satisfactory agreement on the occasional loan. A useful substitute is paper secured on a simple wooden frame, not unlike a target frame for rifle or pistol shooting, with a soft backing such as cardboard or hessian stretched across the framework behind the paper. This gives enough support to stop the paper tearing and therefore gives good definition of the strike of the pellets while, at the same time, allowing adequate penetration.

The frame needs to be at least 5ft (1520mm) square, which usually means gluing or taping together pieces of wallpaper to achieve the required size. As it is being used simply to evaluate shot patterns, and not the other functions for which the larger pattern plate may be used, it is useful to have an aiming mark in the middle of the paper. This is partly because it is the minimum practicable size so you need to be 'on target'; it also allows, when testing a double gun, to check whether both barrels are shooting to the same point of aim, and whether any particular gun tends to shoot high or low. The paper has to be changed for each shot, which is a bit of a nuisance, but at least it gives a good record. On a pattern plate if a record is needed then a digital camera is as good a tool as any.

Results The great thing about patterning is the degree of choke is determined by actual result, in other words by the percentage of pellets for a given shot load that appear within a 30in (760mm) circle at 40 yards/m. This is time-consuming and, therefore, expensive, but it is the only way to regulate the chokes properly; many customers will opt for the more economic option of opening out the choke to a calculated size.

At the same time, from a gunsmith's point of view, it is very satisfying patterning a gun. Results are not always as expected but at least it gives the opportunity of learning a little more each time about bores, chokes and cartridges.

Full choke	75 per cent of pellets in 30in circle
Three-quarter choke	65 per cent in 30in circle
Half choke	60 per cent in 30in circle
Quarter choke	55 per cent in 30in circle
Improved cylinder	50 per cent in 30in circle
Cylinder	40 per cent in 30in circle

A couple of tips:

1 Prior to starting patterning, check the number of pellets for a given cartridge against the manufacturer's specification so it becomes obvious what numbers of shot a certain percentage represents. For example, the English load of 1oz of the popular number six shot size contains 270 pellets; therefore with improved cylinder, you would expect to find about 135 pellets within the 30in circle.
2 It may seem obvious, but it is not worth counting each individual pellet strike; instead, use a marker to identify, say, five pellet strikes at a time then add up the marks and multiply by five for the total.

Some Problems with Patterning The percentage of pellets in the 30in (760mm) circle assumes an even distribution of shot. This neat theory is rather upset by bunched patterns, where the distribution of shot varies considerably, and blown patterns, where there are areas with few, if any, pellets. Often, bunching and blown patterns occur together and it is possible to see patterns with gaps big enough to miss the vital areas of the quarry, as well as dense bunches of pellets almost heavy enough to produce pheasant pâté.

The causes and rectification of this would almost fill a book on their own, but as a guide the following will all have an effect on poor shot patterns:

• the cartridge – most modern cartridges are fairly good but it is always worth trying another brand and even variations in load or wad type. Quite often it is useful to try softer 'game' loadings;
• the choke section – check for finish, shape and length. Too long a choke is not good. This occurs usually where a full choke has been opened out to a much more open choke, which should ideally have, in comparison, a shorter parallel section to the choke. Concentricity of the choke to the bore is also most important;
• the barrels – if the bores are too large, with some loads proper sealing can be a problem. Look also for forcing cones that have been altered but not properly finished, leaving a rough surface that abrades the outside of the shot column, deforming the pellets.

Altering Chokes

Opening out or relieving chokes can be done using a variety of methods. Spill boring just the choke section is fiddly, time-consuming and, unless you

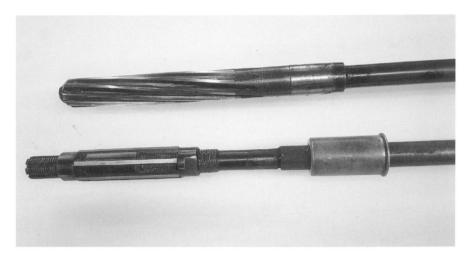

The adjustable reamer on extension has its uses but is a crude tool compared with a properly modified machine reamer.

are doing it just for the experience, not worth the time or the effort. A number of keen amateur gunsmiths who do not want to go to the expense of special tooling use adjustable reamers. A few make the initial mistake of entering a reamer via the muzzle and, after a few chokes have been chewed up, discover what any experienced gunsmith knows: that choke alterations are done from the breech end. With an extension fitted to run true with the reamer and a guide that sits in the chamber it is possible, with care, to remove metal from the choke area.

Once the reamer is set to cut, contacting the choke and set about 0.002in (0.05mm) larger and lubricated, it is turned with a large tap wrench or a brace. The reamer is run completely through the choke, then reamer and extension are removed from the muzzle end. This process is repeated until the choke is within 0.005/0.006in (approx 0.15mm) of the finished size. All metal cutting leaves marks and sometimes minor tears in the surface. Straight-flute adjustable reamers used in thin-walled barrels seem to give a particularly poor finish. The material left in the bore has to be honed or lapped and then polished for the final finish. When lapping it is not desirable to have any more than 0.006in (0.15mm) maximum to remove, otherwise the lapping action produces a slight bell-mouthed muzzle, an altogether undesirable feature.

By far the best way to relieve chokes is by using specially modified spiral-flute machine reamers and doing the job set up in the lathe. An engineer will tell you that the job of a reamer is to finish a bore and normally you would only consider removing a maximum of about 0.004in (0.1mm), but with reamers modified by a specialist tool-

grinding company (see below), it is possible to take out up to 0.010in (0.25mm) in one pass, with good results.

Sometimes in a traditional gunsmith's workshop there will be hanging on the wall an old dust-covered gun stock with a figure of eight cut out intersected by a slot. This would once have been used to hold the breech end of a pair of barrels while reaming out the chokes. The reamer on its extension revolved in the lathe and the barrels were run on to this by hand, the modified stock holding the breech and used to prevent rotation. The idea was to give a good indication of how the reamer was cutting and this was indeed possible, but it could also produce damaged fingers if something went wrong. While it is very much a traditional method, it is definitely one to be avoided.

Better options are to use the lathe with an adaptor in place of the tool post that allows some movement of the barrels, or to have the extended reamer set up in the spill-boring machine. It depends what is to hand. There is no real advantage to either except with my spill borer the carriage is pushed by hand, which helps retain that vital element of 'feel'. It is necessary to choose a reamer pilot that will just enter the existing choke and a major diameter 0.003/0.004in (0.1mm) smaller than the finished choke size. If a lot of choke is to be removed, this often means going up in increments with several reamers. One advantage of this method using spiral-flute machine reamers is a reasonable finish; the final cut is closer to the required finish size than with a straight-flute adjustable reamer.

With everything set up in the lathe or borer, and the reamer pilot engaged but the cutting edge just backed off from the existing choke and well

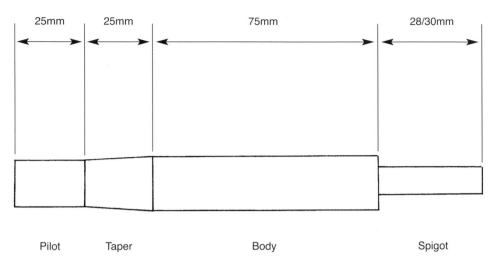

25mm	25mm	75mm	28/30mm
Pilot	Taper	Body	Spigot

Basic reamer drawing. For sizes see the table below.

lubricated with tallow, the speed is set at about 35 rpm and the barrels gently fed on to the tool (by hand wheel when using a lathe). The reamer is fed right through to clear the muzzle and is not under any circumstances withdrawn right back through the choke.

When the required size is achieved it is a matter of cleaning out any swarf from the bore, lapping or honing and polishing the choke section, and just 'softening' the sharp edge left at the muzzle. For this I use a small hand tool known as a 'burr quick' (American), which can be controlled to give the finest of slightly chamfered edges.

USEFUL REAMER SIZES: 12-BORE (METRIC CONVERSIONS NOMINAL)

Pilot (a)	Body (b)
0.679in (17.25mm)	0.689in (17.5mm)
0.689in (17.5mm)	0.699in (17.75mm)
0.699in (17.75mm)	0.709in (18.0mm)
0.709in (18.0mm)	0.719in (18.25mm)
0.719in (18.25mm)	0.729in (18.5mm)
0.729in (18.5mm)	0.739in (18.75mm)
0.739in (18.75mm)	0.749in (19.0mm)
0.749in (19.0mm)	0.759in (19.25mm)

Chrome Bores
One thing you cannot do is ream out chromed bores, which are really hard and will simply take the edge off a normal steel tool or cause it to jam and break. There are fairly expensive tungsten carbide tools available but it is easy to remove the hard chrome with a lap or hone prior to reaming.

Lapping

Lapping is an ancient process that has been practised for centuries either to produce a smooth bore or to mate two components together. It has also been described as lead polishing and, in the context of gun-barrel work, was at one time called draw boring. At a time when technology is seemingly king, the lapping process provides a useful lesson in simple brilliance. For gun-barrel work, a lead lap is used, liberally smothered in a mixture of flour emery and oil. This lap is also bulged in the mid-section giving an approximation of the shape of a barrel (a wooden brandy barrel, that is, not gun barrel). This shape is an aid to producing a parallel bore while a parallel lap will inevitably produce a bell-mouthed bore.

Lapping in many ways does the same job as honing without the capital outlay, special stones and equipment involved with the latter process. If it has any drawbacks, it is marginally messier than honing, neither does it have the range of adjustment available with a hone, but it is a tool that is easy to make in the workshop for little direct cost except an investment of time.

Making a Lead Lap
The material requirements for making a lead lap are modest:

• 1 × length of steel or hard brass rod ⅜in (10mm) diameter × 36in (910mm) long;
• 1 × parallel piece of gun barrel of the appropriate

bore size at least 6½in (165mm) long;
• 1 × sufficient steel or brass to make a plug for one end of the barrel;
• a quantity of lead – the thin roofing lead is best as it is virtually pure;
• flux and solder.

The above, with matching sections of barrel, can be used to produce laps in the size range twenty- to ten-bore inclusive. Rod size for .410 and twenty-eight-bore is ⁵⁄₁₆in (8mm) and, as an aid to rigidity for eight-bore and larger, ⅝in (16mm), which can be in the form of thick-walled tube. If a gun barrel in the required size is not available, steel tube to the next nearest size up can be used, as long as the bore is smooth and parallel and the lap is then turned to finish size in a lathe.

The end plug is turned with a short spigot (approx ¼in (6mm)) to engage the barrel or casting tube and drilled in the centre to accept the end of the rod. Where the lead is to be moulded to the rod, around 7in (180mm) should be cleaned,

fluxed and tinned with solder. As an aid to grip, some cut-outs along this length can be an advantage.

The moulding process is similar in principle to the casting of bullets. In this case the section of barrel stands vertically, with the plug at the base and the lapping rod central down through the barrel section and engaged into the plug. The rod and barrel section are preheated with a propane torch and molten lead poured in until it fills the piece of barrel.

Caution: lead is poisonous and molten lead should not be used in an enclosed environment. Do the work in the open air using a face mask and heavy protective gloves, eye protection and a leather apron. It is also wise to use only the smallest amount of lead that is necessary for the job in hand to reduce any sort of accidental risk to the absolute minimum.

When the mould has cooled to ambient temperature – three-quarters of an hour is a safe period of time to ensure this – the lap can be knocked out from the bottom end of the tube. It may require a trim up in the lathe and at this stage I turn it to a simple barrel shape, tapering down either direction from the centre section.

Using the Lap
Years ago the lap would be set up in a lathe with the tool post moved to one side and the tailstock removed. As with reaming the chokes, the barrels were fed on to the lap breech end first, and the whole process then took place freehand, merrily lapping backwards and forwards semi-rotating the barrels in a kind of figure of eight pattern and providing what any old-timer would call plenty of 'feel'. This archaic but efficient and somewhat exciting process is another that seems to have been contrived to send any health and safety officer into a fit of anxiety. Fortunately, there are similar but safer methods that achieve the same result.

The barrels can be set up on the lathe with an adaptor in place of the tool post in precisely the same manner as for taking out chokes. One alternative is to strap the barrels to a piece of flat wood clamped into something secure like a vice or hobbyist's 'workmate' portable bench using an electric hand drill to power the lap. There is one drawback with this method: the amount of power needed to drive a lap of these sizes requires not only a large drill but smaller amounts lapped out with each setting than can be achieved with a lathe. The advantage of the drill is that it is quicker and more

Cutaway of lap mould. When cooled the end plug is removed and the lap easily tapped out.

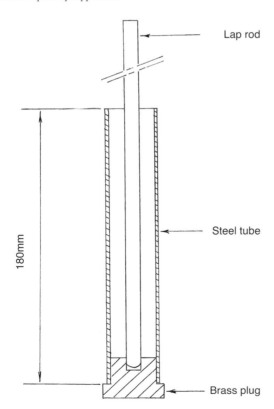

Lap rod

180mm

Steel tube

Brass plug

Lap with collar that fits into the chamber so the rod does not rub against the chamber edge or barrel.

flexible in the back and forward motion than the lathe saddle can be operated and provides much more of that important 'feel'. Because there is less control over the alignment when holding a drill by hand it is best to machine a collar that fits into the breech, to provide proper alignment and prevent any chance of the lap rod rubbing up the inside of the barrel.

Once the means of actually carrying out the lapping process has been settled on, it is necessary to match the lap to the choke and initially it should be slightly undersize at the mid- (largest) point. By squeezing this area in the vice, the lap can be altered to provide a modest interference fit in the choke bore. Then a 60/40 mixture of flour emery and oil is brushed on to the lap, a squirt of oil aimed into the choke section via the muzzle, and the process can begin.

If the lap is too large for the choke it will squeal and feel as if it has come up against a dead stop. This can be a problem but only if the lap is forced harder into the choke section, as it will heat up and can seize in the bore. By gently working the lap backwards and forwards until it runs into then through the choke, this problem can be avoided. It is then a matter of working the lap backwards and forwards through the choke section to remove all machine marks and produce the final finished size. Run the lap quite fast, around 600 rpm as a minimum. Reloading the lap with emery and oil every minute is most useful and as the lap wears it needs to be removed and nipped up again in the vice to provide a tight fit in the choke. It is surprising how quickly 0.003in (0.10mm) is accounted for when lapping. It is, therefore, best to keep a close eye on bore size as it is easy to overrun and finish up with a larger size than intended. For a short choke section with a lap that is initially a tight fit, it is not normally necessary to reset the lap to finish the job.

It can be tempting when only, say, 0.010in (0.25mm) of choke is to be removed to consider lapping it all out rather than bothering to set up a reamer to remove around 0.007in (0.20mm) of material. The problem with this is that as the passage backwards and forwards of the lap is not strictly controlled, and it does pass out of the bore with extended lapping, there is a chance that the barrel at the muzzle may finish up slightly bell-mouthed; this is not desirable.

When the lapping is completed, the residue of the lapping medium needs to be cleared out of the bore. The old-fashioned tow or loose hemp bound on a cleaning rod is one of the best cleaning materials for this job. Then, for a really professional job, it is just a matter of polishing the choke bore!

Bore Polishing

A polishing stick can be made quite easily from a length of rod (the same as used for making a lap), but with a hacksaw slot about 1½in (40mm) down the centreline at one end. This is used to secure the end of a winding of wire wool, '0' or '00' grade, which, when lubricated with paraffin, is used just like the lap in a backward and forward motion running at 600 rpm minimum.

Use of a polishing stick is not just confined to final finishing after lapping. When a bore in general appears grubby, a polish will sometimes be very beneficial.

CHAPTER 6

Advanced Barrel Work

Introduction

Barrels that require a serious amount of work are a job that most gunsmiths approach with at least a modest degree of trepidation. When barrels have reached a stage at which they perhaps have loose ribs and are out of proof and pitted, or just heavily pitted but deemed possibly dangerous to use, they are technically scrap. However, it may be possible to carry out repair work to salvage them and, if necessary, to re-proof and thereby extend their working life. With experience, the skill can be gained to examine and reject barrels that are unlikely to clean up and be acceptable for proof, although it may be necessary to carry out some preparatory work to get a better idea of their real condition.

Occasionally a barrel will clean up and remain within its original proof size but many will require re-proof, which is always a bit of an unknown factor. Imagine spending hours of working time lifting dents, replacing ribs, boring and honing, and putting an action back on the face, for a barrel to bulge when submitted for proof. It can mean a lot of work for little or no reward.

Types of Barrel

It is tempting to think of barrels as falling into two distinct types: Damascus and steel. Some would regard modern steel barrels with chromed bores, typified by the Beretta and the much underrated Baikal, as different again, but the gunsmith's point of view is usually drawn to one aspect: what are they like to work on?

Many barrels when they appear for restoration are quite old. Most Damascus barrels are found on guns produced prior to 1914, for example, meaning that any that come in for work are already over 90 years old. If the gun is an English muzzle loader, you can add on at least another fifty years. Damas-

cus, though, is a 'kind' material, easy to bore and strike up, and much stronger than is generally realized. It is true that it dents easily – and this is one of the most common complaints about it – but, because Damascus barrels are made in a manner that resembles a compacted helical spring, when subjected to damaging internal pressures, they show amazing strength. When they do rupture it is usually confined to a localized section between the original forge welding.

Some of the early steel barrels well pre-date the last Damascus barrels produced in around 1912, and can be very variable in quality. It can be difficult to determine exactly how early they are, as the only guide to age may be the proof marks; even then the tubes could have been around for some time prior to being used to make up barrels. What does characterize these barrels is that they can be most treacherous things, harsh and hard to work. It is quite typical to find a patch of rust that has been hidden under the edge of a forend. Cleaned up, it is found to be a pit, which then turns into a kind of wormhole that follows a fault (or 'grey') deep into the barrel wall. If the barrel is bored, the chances are

A ruptured barrel – an early steel barrel from a twenty-bore single-barrel gun.

that the tool will find hard spots that sometimes tear away, creating further damage. When that type of barrel fails, it is usually in a quite dramatic fashion, with longitudinal splits, sometimes opening up like a ruptured bean can. Even a modest overload can leave it bulged beyond repair and there can be little doubt that some eventually suffer from metal fatigue.

Fortunately, there seem few to be guns left in circulation with these poor barrels and, in the light of history, perhaps they should not be judged too harshly. They were once part of a new technology at a time when Damascus barrels were at the peak of perfection. Steel barrels were undisputedly the way forward, and while some makers got it right very quickly, others did not; everyone got there eventually, and if we did still have to hand-forge all our barrel tubes the modern gun industry as we know it would not exist.

Spill Boring

When I first learnt about spill boring I was told that it was the kind of job where you needed to keep your fingers crossed. Since then, nothing has happened to suggest that this advice was not accurate. When a spill-boring job is successfully completed,

making a previously rust-scabbed bore beautifully clean, it is immensely satisfying. However, it can so easily go wrong and, most frustratingly, this is usually through no fault of the operator. There are a number of precautions that can be taken to minimize the risk of accidental damage.

The main elements of spill boring are the boring bar, the spill, paper packers and the lubricant. The actual bit or tool end of the boring bar can be made from an old square-section file. It is important that, when ground down to form the bit, the sides are both square and parallel. I like to produce mine so they are rectangular in section and the wider side becomes the base that sits against the packing. Two diagonally opposite corners are left square and the other two are rounded off to form a rubbing edge. This gives a bit that can be turned through 180 degrees to use a new cutting edge prior to resharpening.

The most useful way of joining the bit to the bar is with a silver-soldered shallow-angled scarf joint. Using a sleeve to joint the two parts is an attractive idea but sometimes the sleeve can foul the bore and so the joining spigot and sleeve must be reduced in size. Also when joining bit and bar they should be clamped to a suitable length of steel and set up to run as true as possible. The bar needs to be long enough to bore out right through the muzzle end,

The essentials of spill boring. Spill-boring bit made from an old square file brazed to a steel rod, paper liners and softwood spill.

if this is necessary. Around 10in (250mm) is the minimum acceptable length for the bit, although longer is better. A bar of at least ⅜in (10mm) diameter and 36in (around one metre) length is sufficient for most barrels.

The spill is the piece of shaped wood that runs in the barrel and supports the paper packing and the bit. It is sometimes described as the 'shoe', while the paper packing strips are also sometimes known as the 'spills'. (Certainly, the long thin strips of newspaper, particularly if they are folded, do resemble the paper spills that often used to be kept in an old brass shell case by the fireside. They would be lit from the house fire for any number of purposes to avoid wasting matches.) I think it is best to stick with the former terminology as this is probably the most widely used and recognized in the gun trade.

The spill is the length of the bit with a flat side and the other side rounded to match approximately the bore of the barrel, just in section like a segment of a circle with the tangent as the flat side. Soft wood for the spill is undoubtedly best as it quickly wears and deforms to suit any barrel bore.

In a fit of enthusiasm I once decided to make precision hardwood spills, accurately turned and sectioned to match a range of bore sizes. The wood I chose was greenheart, strong, tough and very hard-wearing – a preferred wood for pier legs and mooring cappings, being almost invulnerable to water attack by marine pests. It has its uses, but not for spill making. The result was a disaster. With any reasonable pressure on the bit the greenheart spill would tend to drag in the bore and once, when it snagged badly, the bit was actually torn loose from the bar. Since then I have stuck with pine.

The paper packers, known as liners, should be cut no wider or longer than the spill, otherwise they can get caught up in the bore and try to go around on their own. Newspaper is the traditional material from which to make the liners and it is advisable to stay with the same paper, as there can be a variation in thickness between different publications! Producing accurately matching liners is greatly simplified by the use of a guillotine. In addition to the paper liners, card of a thickness equivalent to usually ten pieces of paper is a substitute when the paper liners build up in quantity.

The traditional lubricant on the bit was a thick smear of tallow along the cutting edge. Tallow gradually melts as the bore warms with the action of the bit and provides limited lubrication over the full length of operation. Webley & Scott used to have a spill borer set-up, where suds oil was pumped through the bore. This lubricated, cooled and,

probably most importantly, flushed away the tiny particles of steel swarf and sometimes rust from inside the barrel. This, I think, is a great part of the secret of a successful spill-boring operation. Using just tallow some of the metal particles are trapped in the residual tallow clinging to the bit while the rest revolves through the bore, pushed along by the bit and spill – a potential problem waiting to happen.

If pumped lubrication is not possible the alternative is to use a larger quantity of tallow-based lubricant that traps the swarf against the bit, preventing it circulating loosely through the barrel bore. A mixture of approximately 40/60 tallow and vegetable oil produces a suitably slimy lubricant, which can be painted on the bit with a small stiff brush and it is pretty well guaranteed to trap most of the swarf.

The Boring Machine

I use a gunmaker-built boring machine which, while appearing more Heath Robinson than Joe Manton, works well. One of the most important aspects of the design (although Alan Preston, the maker, never 'designed' things; he simply had an idea and got on and made it with whatever was to hand) is the flexibility of adjustment for setting up the barrels. It ensures alignment with the boring bit and allows some movement while the job is in progress, which is important. Speed is around 40 rpm, which drops slightly under load; a simple friction-clutch arrangement releases the drive if, for some reason, the bit tries to jam or snag.

The saddle or carriage that carries the barrels has no power drive but is simply pushed backwards and forwards. Originally this was linked to a system of weights running over pulleys to reduce the amount of arm work.

There are proprietary spill-boring machines around, such as the double-headed Holroyd, which was favoured by BSA when they produced shotguns. Unfortunately, with few purchasers left for spill borers, such machines are rarely seen – probably a lot have been scrapped – although some have been converted to deep-hole borers. The problem with these proprietary machines is that they are usually too bulky for the single-handed gunsmith, who has limited space.

A lathe can be used as a borer by removing the tool post and fitting a simple adaptor, which holds the barrels but allows some movement like on the spill borer. This flexibility of movement is necessary because, as the packing builds up under the bit, it is pushed further off-centre and, instead of

More Heath Robinson than Joe Manton – my gunmaker-made spill-boring machine set up for demonstration purposes, hence guards removed.

revolving about its axis, actually describes a circular path and imparts a waggling motion to the protruding barrels.

Boring a Barrel

The bit is set as viewed from above with the rounded rubbing edge as the top leading edge closest to the operator, to rotate anti-clockwise when viewed from the bed end of the machine, just the same as the lathe set in forward motion. Initial packing between the bit and spill with the paper liners should be done with the boring bar detached from the machine drive. This is the safe way of setting the initial packing to avoid the possibility of an early seizure due to over-tight packing. On the other hand, it is necessary to avoid any looseness in the packing as this can allow the bit to revolve against the liners and spill, chewing them up and then clattering around the bore with the possibility of marking the bore.

When making the initial setting it is normal to put in enough paper liners so that the bit, liners and spill, when pushed past the chamber and into the bore, fit together snugly and can just be revolved by hand. This is usually the slackest part of the bore so as it works up the barrel it will tighten, but not normally enough to cause any problems. If there are enough paper liners to exceed the thickness of a card liner it is best to substitute card for the equivalent thickness of papers as this seems a help in reducing vibration.

It is most important when boring the main part of the barrel not to run the bit into the choke section unless it is cylinder bored. Obviously, the choke section is tighter than the bore, sometimes considerably reduced in diameter, and this would jam the bit and result in the possibility of irreparable damage. Before proceeding it is vital to identify the depth to which the boring bar needs to engage.

The start of the choke – the cone from the breech end of the barrels – can be determined by measuring from the muzzle with a bore comparator, or from the breech with a plug that is a sliding fit in the main bore and a steel measuring tape. Once this position has been identified, a corresponding mark should be made on the outside of the barrels and measured against the boring bar. It is tempting simply to put a mark on the bar or wrap tape around, but this is not sufficiently accurate and a flat bar clamped on to or across the bed of the machine to act as a stop is the best method to limit precisely the depth of engagement.

With everything set, it is, perhaps, necessary to point out as a final check for safety reasons. That is, the barrel to be bored must be dent- and bulge-free, leaving only pitting or rusting to be bored out; it must not be a chromed bore; and the extractor(s) must be removed so they do not snag on the boring bit. Once you satisfied that everything conforms, the barrels may be mounted in the barrel clamp or adaptor, the tail end of the boring bar fitted into the borer chuck or driving head, and the lubricated bit, liners and spill just entered into the actual bore by about 2in (50mm). This is done by moving the carriage forward and the easiest way to hold the bit, liners and spill together while this is being done is with a ring cut from a short length of copper tube; this can be slid back out of the way when no longer needed.

This is the moment of truth. With the guard in place, once the drive is switched on the barrel should be moving across the bit and not dwelling in one place for more than two or three seconds. With a spill borer using manual movement of the saddle, the user gets a lot of 'feedback' and it quickly becomes easy to detect heavy rusting, tight and slack sections of barrel, and places where dents have been lifted but the bore is not quite perfectly round. Using a lathe, the initial work at least should be done manually using the wheel on the saddle. As with the manual spill borer, initially it allows the feed rate to be variable so when a tight spot is found the feed can be less than on a more open section of barrel. Once the barrel bore is trued up – and this is the great benefit of using the spill borer – the feed rate can be constant, at about ³⁄₃₂in per revolution. It will then take about three and a half minutes to bore a 30in barrel less the choke section.

The action of the bit is not to cut but to scrape the bore. The swarf produced is very fine, resembling steel (and rust) dust. It is not a quick process and certainly not for the impatient. Each pass will remove only about 0.0002in (0.005mm) from the bore, although where there is rust scabbing early results can be seen quite quickly.

Withdrawing the bit is done while it is still rotating; this should not be rushed, but it is obviously a lot quicker than when feeding in for boring. Getting back to the starting point is simply a matter of stopping the borer and running the barrels off the bit while holding spill and liner in place. The tool is cleaned and re-lubricated, a cleaning rod pushed through the barrel bore to remove any loose swarf and, with another liner inserted into the packing, the process is repeated.

It is worth, about every fifth cut, doing a spring cut – a pass up the bore without adding any extra packing. You should also examine the spill to ensure there are no sharp particles of metal embodied in it, which might mark the bore.

Some Problems with Spill Boring

Vibration can set up a pattern of longitudinal chatter lines along the bore, which become impossible to bore out. It then becomes necessary to remove them, preferably by lapping, as such work can cause undue wear on, or even break, honing stones.

The potential for chatter lines can be inadvertently produced if the initial packing is too tight and the bit rammed into the barrel leaving a fine line cut into the bore by the cutting edge of the bit. Usually, however, this is not a problem. It is only noticed as a clicking noise as the rotating bit trips over it, and will normally clean up.

There are other times, either when cutting or withdrawing the bit, when a sharp 'dink' sound is audible. This usually indicates that a piece of metal is picking up in the bore and executing a deep spiral groove, which can ruin a barrel, usually before there is a chance to hit the stop button. The reason why this happens is often a mystery and, equally, deeply frustrating. It is rarely possible to reclaim the barrel, as the damage is often too deep. If the barrels are a little on the thin side, it may even push through to be visible from the outside.

Even with care and a very fine cut, spill boring does leave marks in the bore. These are seen as fine concentric rings, which have to be lapped or honed

Sometimes things can go badly wrong – the damage marks where the tool has picked up are even visible from the outside.

out as the final part of the finishing process, so it is best to finish boring about 0.002in (0.05mm) below the finished size to allow for cleaning up.

Why do we still indulge in the somewhat incredible process of spill boring? Well, simply because it has its uses. Pits and rust scabbing may tear a honing stone to pieces, but the boring bit will cut them out with ease. If there are undulations in a bore, the long boring bit will true them up producing a good straight-sided, albeit slightly tapered, bore.

Spill boring is an ancient technique practised today by only a few, but this sort of knowledge and these skills should not be consigned to the dustbin of history just because there may be something better. The hone, in spite of its limitations, has taken on much of the role once exclusive to the spill borer and has proven to be a good halfway house between the lap and the borer.

Honing – The Machine

It is not too difficult to find small second-hand horizontal honing machines at prices that make it hardly worthwhile either building a dedicated machine or converting a lathe. One of the well-known makes is Delapena of Cheltenham, Gloucestershire, whose much earlier but identical machines appear under the Delson name. The major cost of setting up a honer is often in obtaining the correct mandrel that carries the stone, along with the wedge that lifts the stone. Both items have to be lengthened to be suitable for work on shotgun barrels, which will of course involve extra cost, but these purpose-made machines are convenient and easy to use. They also have adjustment that can be 'dialled in' while you work, although the graduations on the dial appear to be a guide only and do not directly relate to a measurement at the stone.

Some gunsmiths make their own honing head holding up to three stones (in honing terminology a single stone is held in a mandrel, while more than one is held in a 'head'). While this works more quickly, the drawback is usually a screw adjuster at the end of the head requiring the machine to be stopped each time for readjustment. In practice, the single-stone set-up with a careful choice of stones seems to work quite well. For barrel work, Delapena grades C6C (fine) and C3C (coarse) seem a good choice.

Evening homework in the garden: Delson hone being built on to a wooden bed with extended mandrel for barrel honing. The bed and carriage are being made from greenheart, which is particularly strong and stable for this type of application.

With any of these horizontal machines operating over a reasonable length of travel, lubricating the stone can be a bit of a problem. Ideally the honing fluid is pumped into the bore so that it not only lubricates and assists the honing action of the stone, but also flushes the bore clean. With a vertical machine, where the job is fixed and the honing head moves up and down above it, this is easily accomplished, but it is more difficult with a horizontal machine where the job moves backwards and forwards. I have seen every conceivable sort of set-up, from a machine with a long trough underneath and honing fluid pumped through the barrel from one end through a flexible pipe, to the chap who feeds the barrels on to the stone(s) by hand with a squirt of oil from a can.

I like the idea of a drip-feed rather like the set-up used on Greeners' old rifling machine. This wonderful and very early machine, later restored by Duncan Bedford of Nuneaton and now at the Royal Armouries Museum, was originally set up at a slight angle so drip-fed lubricant would dribble along the bore. Duncan once told me that according to one of the original operators, this 'fall' on the bed was achieved by cutting out floorboards at one end and standing the legs on the joist underneath. This story has a ring of truth that sounds about right for the gun trade, so if it was good enough for Greener's…

The lubricant used for many years was commonly known as 'red cutting oil' and supplies occasionally surface at second-hand machine-tool sales. New supplies are no longer available commercially because it is now regarded as potentially damaging for health reasons, and very bad for someone suffering with a skin complaint such as eczema. When the oil was in regular use, operators would be issued with long rubber gloves and rubber aprons, but a peculiar biting smell would still impregnate their clothing and skin, despite all precautions.

It is tempting to make your own honing fluid and plenty of users have done just that. A mixture of light-grade car engine oil and paraffin used to be a favourite and even neat diesel has been tried. Unfortunately, while economically this is an attractive proposition, we are now informed that even a paraffin mix is unacceptable due to the – albeit very slight – possibility of damaging the skin. So it is a case of purchasing commercial honing fluid, which, of course, all the old hands will swear is not a patch on the stuff they used to have!

Honing Versus Lapping

There are obviously many similarities between lapping and honing as it is essentially the same process. The great advantage of honing is the ease with which the adjustment of the stones can be made and it is also a cleaner process, especially when examining the bores while working on them. A couple of cloth patches pushed through will leave the honed bore clean for visual examination, while lapping will require tow and patches applied sometimes eight to ten times to clean the bore. Honing is also a fairly efficient way of removing small pits while similar work with a lap can be very time-consuming. Also the hone is at an advantage when using coarser stones. If a grade of emery grit much coarser than flour emery is used to speed up the lapping process, the main effect is to wear away the lap at a rapid rate until it becomes too slack in the bore to be effective.

Proprietary honing mandrel. This type has a single stone with shoes opposite giving three-point contact. The extension to the mandrel needs to be at least 26in (660mm) long.

When fine honing, equally good results can be obtained with the lap and it is quick and convenient to change to the finer grades of lapping paste as the work progresses. The lap costs hardly anything to make and when not in use can stand in the corner or, better still, be hung on a board. The hone costs money – a set of six stones (the minimum order quantity) at present costs around £35 – and the whole thing, even if in take-down form, uses up valuable workshop space, which may be at a premium.

The hone and the lap have complementary uses. For small jobs and odd bore sizes, the lap is more convenient and has a flexibility of use that the hone lacks, without making or investing in an extensive range of tooling. Set up for common bore sizes such as twelve- and twenty-bore, the hone is very efficient; then, the only time the lap needs to be brought into play is when a particularly tight choke is outside the adjustment range of the honing mandrel.

Honing a Bore

The horizontal honing machine is similar to the spill borer in that the barrel(s) can be set up, less extractors, on a carriage that is moved backwards and forwards by hand, but at a quicker rate. As the bronze shoes at the back of the mandrel opposite the stone are in constant contact with the bore, and the driving tube aligns with these, there is not so much offset or accumulative misalignment of the bar as they may be with spill boring. Therefore, it is not as important to have quite such a flexible clamping arrangement; shaped hardwood barrel clamps lined with felt are sufficient. If a three-stone head is used, the alignment remains truly central in the bore at all times, as long as the stones are of matching thickness. If stones have been previously used, they should be inspected for small particles of metal that become embedded in the surface. These should be picked out prior to use.

Just as with spill boring, when lapping the main bore it is necessary to measure the depth of engagement of the stones to avoid running into the choke section. Once again, fixing some kind of strap on the bed of the machine is the best way to achieve a positive stop. Personally, with honing I like a stop at both ends of the travel, so there is no danger of pulling the stone into the chamber. If this were to happen, it could come loose, jam and break. If only the choke section is to be worked, then, as the stone is at least over three times the length of the choke, it is easy to work it without resorting to carriage stops.

With the honing mandrel entered into the bore in front of the forcing cone it should be tightened so it will make contact for most of the length of the bore. There may be some variation here due to taper, undulations caused by previous bore work or small bore defects remaining from dent lifting. Once the initial fit has been arrived at, it is a matter of applying the honing fluid by the chosen means, switching on the machine and working quite smoothly back and forth between the stops. Try to avoid dwelling in one position, particularly at the end of the travel where the motion should be reversed quickly and immediately to avoid a bowed or belled bore.

Take note, though: small proprietary machines are often operated via a foot pedal that releases the brake and tensions the driving belt to run the machine. Operation of the foot pedal also exerts some pressure on the wedge to lift the stone, so ini-tial setting must be done with the motor off and the foot pedal depressed to pre-tension the fit of the stone, otherwise it will be much too tight in the bore.

At first with an imperfect bore the adjustment of the stone will be fairly frequent. As the bore trues up, or the pitting is reduced, adjustment will occur less often, as long as care is taken to apply plenty of honing fluid to avoid undue wear on the stone. It is always tempting to work just one part of the bore where imperfections have been identified, but this should never be done. The whole length of bore between forcing cone and choke cone should be worked evenly, to avoid changes of bore diameter. There is also a natural tendency to hurry the process by a little extra tightness on the stone set-ting, but this should be resisted; the stone should make sufficiently firm contact to avoid chattering but should not be so tight that the passage of the stone up the bore is obvious from the outside. To a certain extent, with an increasing familiarity with the process, and with soft or thinnish barrels, it can be possible to 'feel' the progress of the stone, as indeed it is with the lap. However, if it is over-tight, there will be a tendency to seize and snap. With a conventional side-by-side, it may also tend to ride up and over the spacers laid between the barrels with the possibility of leaving an undulating bore, which should be avoided. One gunsmith I knew (who has long since gone to the great gunmaker's in the sky) used to strip off the ribs and remove the spacers if the barrels were on the thin side, to ensure a concentric and parallel bore. It is a lot of work but he was a brilliant barrel man and I saw some he had got through re-proof that seemed almost paper-thin towards the muzzles.

Which brings us to the question – what exactly is 'thin'?

It is generally accepted that 0.030in (0.75mm) is a good practical minimum to allow for a little work and a good chance of passing re-proof. Taking a look at this measurement in a set of mechanic's feeler gauges really brings home the fact that it involves not very much metal at all. The only rea-son it has any strength is because it is part of a taper walled tube.

Hopefully, the honing process will go well and a clean bore will be the result, after a final hone with the fine stone, and polishing with the polishing stick and wire wool. But do not be too hasty – it does not always work out that easily.

Problems Caused by Honing

Many barrels have been spoilt by honing, some by

unscrupulous practice, some unintentionally, due to a lack of knowledge or skill. A shotgun is required to pattern properly and, if it is a good-quality gun in original condition, it is certain that someone has spent a lot of time boring and regulating the barrels to shoot well. Being, in the main, obsessed with shiny bores, shotgunners will go to great lengths to have any blemish removed, even those in the strongest parts of a barrel well in proof, but will the gun pattern the same after honing? It is a sobering thought that, for the sake of a few small marks that could have been kept under control with regular and proper cleaning, all the good original work may be undone.

The other aspect that is common to all barrel work is the effect it has on choke. Obviously, if the main bore is opened out and no work is done on the choke section, the effect is to increase the choke. It follows, for example, that a main bore opened out 0.010in (0.25mm) in diameter will increase the choke as a comparative measurement

by the same amount. It is something to be noted and, while if may be necessary to open out the choke a little to compensate, there are other times when it can be advantageous.

Re-Laying Ribs

The re-laying of ribs is a job redolent of alchemy and mystery; hot irons, fumes of solder, the sweetly pungent smell of burning resin, and an atmosphere seemingly filled with an ethereal smoky haze. Since the long-gone days of soldered pots and pans, and lead water pipes, traditional soldering methods have gradually become little used, apart from a blob or two on a circuit board. The skill is in danger of becoming lost and, as a result, shrouded in mystery. Yet in reality there is no mystery at all, just careful preparation, the application of delightfully simple technology, and, of course, practice.

A pair of barrels wired and held in the modified rifle cradle. This gives minimal contact with the barrels, which means that heat is not drawn away when re-laying ribs.

Preparation

When a rib becomes loose it is quite often only one rib and even then only part of that same rib. With the exception of the top sighting rib on an over-and-under, it is best to strip off the defective rib and the rib or rib/forend loop/keel plate/keel rib on the opposite side to allow a full examination and subsequent preparation. With a side-by-side, it is more often the top rib that has come loose – those smaller ribs under the barrel seem to cause far less problems – but in re-laying, a proper job involves doing them all together, so that it is known to be a sound assembly. It is extremely frustrating, after re-laying one rib, to find the unattended rib 'spewing' from a pinhole (initially bubbling when in the blacking water, and then 'spewing out' water that has entered through the same small gap), and spoiling the chances of making a good job of the barrel blacking.

The ribs, especially if very loose, can be removed by heating the barrels with a propane torch whilst held in a cradle. Care must be taken to wear reasonably heatproof gloves such as heavy leather riggers or hedge-laying gloves. While the intention is always not to touch hot metal, accidents do happen and good gloves can prevent nasty burns. Some years ago a customer kindly gave me an old sighting cradle used for bore sighting Lee Enfield rifles. With its main support legs shortened, this made a useful aid for barrel rib work, for both stripping off and reassembly. Otherwise it is simple enough to make a cradle. The heating should be as even as possible and the ribs allowed to become detached without resorting to prising off with any form of pointed tool. This may bend or kink them, especially bottom ribs, which can be very fragile. An alternative is to use the heated barrel irons to heat up the barrels for rib removal; this may give a more even heating, but it is more a matter of choice than for any real technical advantage.

With the ribs and barrels suitably cooled off it is easy to make a full examination, and if the ribs are not rusted away along the edges, and no problems, such as excess rusting or deep pitting, are detected between the barrels, the first steps towards reassembly can be made.

Surprisingly, the tinning along the barrels is often quite satisfactory even where the rib has pulled away and, of course, there is no rusting in any area where the tinning is sound as the layer of tin or solder protects the steel underneath. Excess solder – in other words, that surplus solder that is more than the thin tinning – should be removed, otherwise it may prevent the rib sitting properly in place

at reassembly. A white metal scraper is a useful tool for removing solder in small areas, or, if there is a lot of it, the barrels or rib can be heated until the solder melts – seen initially as a subtle change of colour – then the excess can be wiped away with wire wool dipped in resin-based flux. A convenient way of holding the wire wool in gloved hands is to bind it to the end of a short hardwood stick.

Areas that are not showing solder where there should be some obviously have to be first cleaned down to bare metal and fluxed before tinning, which, in spite of its name, is actually a layer of solder. Solder paste or paint, a mixture of solder and flux, can be used for this process, applied with a small brush and then heated. Globules of solder appear as it heats up and the flux spreads out, then the solder will melt across the fluxed area, which can once again be spread to the thinnest layer.

The problem with the solder pastes or paints is that they usually have a cleansing flux, which means that they are, in the long term, corrosive. This is why, in spite of the possible convenience, ribs should never be put on with this type of product. It will corrode between the barrels, where it cannot be seen or cleaned away and will have the potential to become a kind of long-term time bomb. However, for tinning operations where it can be neutralized or washed off afterwards with hot, soapy water, to negate the corrosive effect, it is a product that is without equal.

Fitting and Fixing the Ribs

The ribs should, when laid on the barrels, make contact on both sides for their full length. There is no point in just trying to fill across a gap with solder – this is difficult to achieve and, anyway, makes an inferior joint that will only fail again. Fortunately, if the ribs have been taken off carefully and are not bent or kinked, they normally fit well.

There are various ways of holding the ribs in place for soldering but still the most common is to wire them on. With the wire some form of spacer needs to be used and the great standby in these matters was always blacksmith's cut nails, which make very effective taper wedges. An alternative for concave ribs is round bar of a diameter that sits neatly on the rib. The main difference is that the nails can be tapped into place after the wire binding has been fixed, while round bar is secured as the wiring takes place.

Most top ribs are proud of the barrels at least for part of their length so it makes sense to secure this first with a non-corrosive solder flux applied to both barrels and rib. There was a fashion for sunken top ribs, some of which appear as slender as bottom ribs, but they do not appear very often nowadays. The bottom ribs and forend loop are wired on in the same manner. It is beneficial to sprinkle a little powdered resin between the barrels prior to fixing the bottom ribs. Not only is this a flux but, when the job is finished, it sets between the barrels and ribs to the consistency of hard varnish helping to protect against any form of corrosion. It is wise not to use too much; while heating the barrels it will melt and boil up and try to find a way out along the rib, and sometimes leave holes blown through the solder along the joint line.

When everything is wired in place it is time to check that the ribs are lying straight and true, and not further up the side of one barrel than the other. The forend loop should be correctly positioned and it is worth double-checking that it is facing the right way!

Electrically heated irons are a great rarity, and therefore the ultimate luxury, and most gunsmiths have simple old-fashioned irons that have to be pre-heated; with this type, it is best to have four, so while two are in use (and losing heat), the other two are ready to hand when needed. The irons are normally heated on a gas burner and judging the temperature is difficult. If on applying stick solder to an iron it immediately melts and runs, the tem-perature is about right. With the barrels in the cradle a pair of irons are entered simultaneously from the breech end. The first two irons really provide a pre-heat, while the second pair normally do the job.

The tinning of ribs and barrel is not enough to form a good joint so the application of extra solder is needed and everyone has their own favourite methods for this. A fairly easy way of ensuring a supply of solder along the joints is to mix solder filings into the flux along the edge of the ribs. It is a tedious job producing the filings but it does provide a simple and effective means of getting solder in the right place.

Some Problems with Rib-Laying

Initially the tendency is to use far too much solder and fill the area between the barrels and edge of the ribs with a thick fillet. This should be avoided, as it results in a lot of really unnecessary work to remove it and, while the job is in progress, it is not possible to see what is going on, such as the rib lifting in one area and leaving a small gap.

Even with care and marking the position, the forend loop does not often finish up exactly in the place where it was originally fitted. Some work can be necessary to make sure the forend iron fits correctly when the barrels are assembled to the action.

Ideally, barrel irons should be copper or brass but steel also works surprisingly well. For any bore or gauge the major diameter needs to be a little smaller than the smallest bore size and the end turned to fit through the tightest choke.

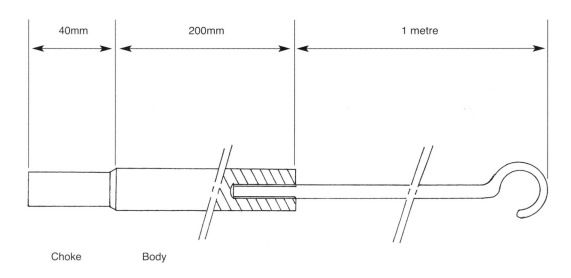

Obviously, any such work is best done while the barrels are still 'in the white'.

When the barrels are hot, the expanding air and sometimes surplus flux have to go somewhere and this may leave a tiny hole in an otherwise good joint. This can be attended to with some solder and a small pencil flame burner on the propane torch while the rest of the assembly is still warm but cooling.

If barrels have at some time been trimmed, and occasionally with some that have not, small gaps may appear at the muzzle end. This does have the advantage of providing a 'relief point' for the hot air and any surplus flux, but it has to be filled before the job can be declared finished. A wedge of solder tapped into place will hold surprisingly well but it should be melted to form a sealed joint and this again can be done with a pencil flame on the propane burner, plus a certain amount of care.

CHAPTER 7

Rejointing Barrels to Actions

Introduction

The hinged, break-open type of shotgun will eventually shoot loose. Correcting this deficiency is referred to as 'rejointing', 'putting back on the face' or simply 'tightening', the latter perhaps suggesting that less scrupulous methods of repairing the action have been used. When the barrels come loose they may be described as 'off the face' or 'having a headache', the latter usually means that they are flopping about like a broken branch in the wind and that there is excessive headspace between the cartridge and action face. Putting the barrels back on the face – meaning the action face – requires the gunsmith to carry out remedial repairs, replacing or building up worn parts, then refitting them by hand.

Shooting loose is a problem that has plagued this type of shotgun from the earliest days of the design.

Subtle variations have been tried with different methods of locking, culminating with the triple-bite system and even varying lengths of action. On a modern mass-produced twelve-bore, the distance between the action face and centre of the cross pin is around 1⅞in (45mm), while some guns, such as the Greener Empire model, have a noticeably longer action bar; (there was also a short action version of this gun). Sometimes these long actions were made to accommodate a particular cocking mechanism and they can present advantages, but such debate is outside the scope of this book. There are often arguments out on the shooting field about whether the long or short action bar is best, but in reality most gun dimensions are a compromise between balance, handling and aesthetics, and the integrity of fitting and quality of materials used are probably more important than subtle dimensional differences.

An unfinished modified Walsh action by Edwinson Green with the cross pin only ½in (12mm) from the action face – an intriguing idea and possibly the only one of its kind. The wide gape makes it easier to load than the bottom barrel of an over-and-under.

Remachining a hook in the miller to remove previous repair of poor quality (machine guards removed for photographic purposes).

The fact that the cross pin is some distance below the centreline of the barrels is a disadvantage, if a practical necessity with side-by-sides. For over-and-unders, setting hinge discs high up in the side wall of the action has theoretical advantages over a conventional hook, and in an ideal world the point of hinging a side-by-side gun would lie along the centreline of the barrels.

One idea on this subject, which achieved a fair degree of promotion but little, if any, commercial success, was put forward in the nineteenth century by J.J. Walsh, editor of *The Field* magazine. Walsh took out patents in 1866 and 1878 for an intriguing design where the cross pin was located at the base of the standing breech, rather like a door butt hinge. It is difficult all these years later to determine whether it may have had some advantages, but presumably it was considered a useful development worth exploring. A later unfinished modified action built by Edwinson Green has survived with the cross pin ½in (12mm) in front of the action face. It has a gape still somewhat more than a modern over-and-under, and the distance between the trigger guard and forend would have been so small as to possibly render it rather oddly balanced to carry while broken over the crook of the arm.

The shotgun world is normally ruled by a good compromise between practicality and aesthetics, with a liberal helping of conservativism, and the 'Field action' as it became commonly, if incorrectly, known never made it much past the starting post.

However, the problem of shooting loose, or more often wearing loose, still exists and is a staple of the gunsmith's work diet.

Methods of Repair

Wear occurs in several places: on the cross pin, the hook, the bites and locking bolt. The parts most susceptible to wear are the hook and cross pin and quite often when these are repaired no further work is required. In an ideal world, and certainly on an expensive gun, the cross pin would always be replaced with an oversize one. This has the advantage that the hook is being refitted to a completely round pin and therefore good contact can be made between the two parts; this is a necessity if reasonable longevity is to be achieved.

It is not always possible to replace the cross pin, as some side-by-sides, including models made by BSA and Webley & Scott, have the 'pin' machined as part of the action bar. On other guns the cross pin is full width of the action and the ends are liberally engraved, and the customer may balk at the total cost of repairs. Other cross pins may be found behind caps (or discs) either screwed or pressed into the action bar (done this way because the pin sits within the action slots to clear the cocking dogs), and these are conveniently replaceable. Even better for convenience are the hinge discs used on some over-and-under shotguns, some of which are pressed in, while others are screwed in.

Repairing the Hook

When, for various reasons, the cross pin is not replaced, the only option is to repair the hook. Various methods of doing this have evolved, some much more desirable than others. One of the oldest types of repair, which can be done at the bench with only hand tools, is to insert new metal into the hook in the form of a dovetail joint. Done by a craftsman, it is visually neat but an imperfect solution as it provides a new bearing surface over only a limited area of contact. It has often, however, been good enough to get a gun through re-proof.

Using more modern technology, similar results can be achieved by welding up part of the hook to provide new metal that can be reshaped and refitted. This usually means cutting away part of the hook prior to welding to form a 'weld preparation', in other words, to ensure that there is a reasonable thickness of weld metal into the hook. Oxyacetylene welding puts in too much heat so is not much use. Electric arc or stick welding is a savage, almost uncouth process to use on a gun, but it has been done and will undoubtedly continue. MIG welding, a similar process to stick welding but using a wire and inert gas, is a little kinder, but not much. There are problems with either of these methods of welding: first, cleaning up the signs of weld deposit; and second, and more concerning, the heat-affected zone two to five millimetres away from the edge of the weld that can cause brittleness in the parent metal.

If welding is a chosen option then TIG welding should be used, but the very best has to be laser welding, which has virtually zero heat input and, when cleaned up, like TIG is almost undetectable.

Some twenty years ago I experimented with a process called spray fusion, which has been used industrially for many years in precision engineering to reclaim worn or mismatched components. The heat input is very modest and the deposited material is laid down literally only in thousandths of an inch. With the correct material sometimes only the slightest difference in colour (usually a creamy appearance) is visible inside the hook. The disadvantage is that many suitable deposits are still hard enough to necessitate the use of a diamond file when rejointing. The main disadvantage from my point of view was having to sub-contract the work to a specialist engineering company some distance away, resulting in extra time and costs (although they were most accommodating and their work was always excellent).

New metal may also be put back into the hook by fitting a half shell of steel. To obtain sufficient wall thickness to make this sort of repair, the existing hook is machined out on the miller to a larger size. To make the half shell a steel tube is turned so the outside diameter matches the milled-out hook and the inside diameter is a little smaller than the original hook dimension. The outside of the steel tube does need to be a good mechanical fit into the oversize hook and this can be checked by smoke blacking. With a small radius along the edge of the hook the tube is silver-soldered in place using a small oxyacetylene nozzle. While some solder will penetrate between the hook and tube, the object is to secure it in place only along the edge, the load being taken between the back of the insert and the hook. Then the tube is cut off until it literally forms a half shell in the hook. Get the inside dimensions just right and it can be a quite modest job to refit the barrels to the action. The giveaway with this repair is the faint gold colour that shows the line of silver solder, although for an average gun it does make a tidy and economic method of repair.

The tool-manufacturing industry uses amazing adhesives to hold tungsten carbide teeth on to circular saws and similar cutting tools. I have experimented with some of them to hold half shells into hooks, so far without consistent success. I still think that there must be an adhesive availabe somewhere that would prove sufficiently strong and durable to make such repairs both possible and virtually invisible.

Replacing the Cross Pin

Where it is possible or acceptable to replace the cross pin it should always be done, as this is by far

Insert silver-soldered into hook. Note the handles of the heat sinks protruding from the breech end of the barrels. The rag wrapped around like a loose bandage is first soaked in cold water.

H.E. Pollard hammer gun with a full-width cross pin held only with a tapered end.

the best method of repair. There can be some complications to take into account, depending upon the type of action. Hammer guns usually have cross pins the full width of the action bar, some are parallel, some stepped, others screwed at one end, held with a locking screw or tapered end. If there is a different way of doing it someone has usually thought of it and sometimes it seems that each gunmaker was determined to 'do his own thing'.

With self-cocking guns utilizing cocking dogs or lifters through the side of the action, the cross pin lies on the inside of this mechanism, which limits the scope for variation. Cross pins may be screwed at one end for location, pressed in and/or held with a locking screw.

Over-and-unders can have cross pins the full width of the action; this is possible because on many the locks are cocked with cocking rods lying under the cross pin. However, for those with ejector trip rods these may come through holes towards the ends of the cross pin. These holes are rarely on centre so a pilot hole and some careful reaming are usually necessary.

The hinge disc may appear to be an insubstan-

Hinge discs removed from a Lincoln over-and-under. Note the location grub screws that need to be unscrewed prior to trying to press out the hinge discs.

tial way of holding a shotgun together but it does work well, possibly because the pivot point is closer to the centreline of the barrels. Where these hinge discs are screwed in – a quite wonderful idea – it is simply a matter of unscrewing them and changing for oversize replacements. When they are pressed in the principle of reaming to fit oversize discs is very similar to replacing a hinge pin.

Reaming for Oversize Pin or Discs With so many size variations it would be uneconomic to have a range of piloted reamers to cover every eventuality – unless you were specializing only in a certain make, in which case it is a fine method of doing the job. The most frequently used tool is an adjustable reamer and three or four sizes are sufficient to cover most of this sort of work.

Once the pin or disc is removed, the reamer can be entered in the now-vacant hole, nipped up until it just bites, then run right through. It is then reset in the hole and the process is repeated. It sounds as if it should not work but it does, and remarkably well at that. It is important where hinge discs are fitted to ensure the reamer is supported through both sides of the action and always worthwhile after a cut reversing the reamer to come back from the other side.

With a pressed-in pin or discs there has to be an interference fit to prevent movement. Realistically, size for size will provide this but having the reamed hole 0.005in (0.125mm) smaller than the mating part will boost the confidence further. When pressing hinge discs into over-and-unders it can be done in a vice, but the thin walls of the side of the action will need to be supported with a tight-fitting hardwood block.

The Bolt and Bites

Noticeable wear on the locking bolt and the mating bites is less common than on the hook and cross pin, but it is none the less an area subject to wear. This is particularly noticeable when barrels have been put back on the face and with the gun shut the top lever lies parallel with, or even to the left of the centre-line of the top strap, when it should lie slightly to the right.

If the locking bolt body has become worn and is sloppy in the action, spray fusion or laser welding along the top face, followed by refitting, is an option. This has the advantage of pushing the bolt further into engagement with the bites. It is worth noting at this point that other methods of welding along one face may well distort the bolt, leaving it an interesting banana shape. The tidiest way is spray

Smoke blacking a hook during rejointing. In the early stages only the hook needs blacking; later the breech ends of the barrels will be blacked as well.

fusion or, for the more adventurous gunsmith or a customer with expensive tastes, making a complete new bolt.

When the fit of the bolt body is deemed satisfactory, it is usual to tighten up the locking by building up the tapered part of the bolt that engage the lumps. This again is an ideal job for spray fusion as it retains the existing angle and can be used to lay down only a thousandth of an inch or so of deposit, which is often quite enough to do the job. Some refitting is always necessary and this is done by smoke blacking the bolt with it detached from the spindle or top lever, so it can be just tapped into place to locate the barrels, thereby simulating the pressure exerted by the top lever spring.

Refitting Barrels to Action

When an oversize cross pin has been fitted – of a suitable size to allow for wear of the original pin and wear in the hook – the barrels will no longer fit on to the action. The same applies when the hook has been built up or had some form of insert fitted. This is where smoke blacking ('blacking down' in gunsmith's language) the barrels and action, so often the centre-piece photograph of magazine articles, is used to great effect.

My old mentor used to tell me that there is nothing thinner than smoke and, while this does not sound very scientific, he was probably right. However, what we are dealing with is the deposits left by smoke or incomplete combustion, a thin layer of carbon. These thin layers can quickly be built up into much thicker layers by progressive application, so it is worth making the point early on that to be really effective the surplus evidence of each previous smoking should be wiped away prior to re-smoking the mating parts. If this is not done, there is a danger of compromising the integrity of the fit of the mating parts.

At the start it is worth checking the action face, where burrs may be located along the edges, mostly with elderly guns. These have to be stoned flat before attempting to fit the barrels.

To start with, where a new pin or half shell has been fitted, only the leading edges of the hook will engage the cross pin. The breech end of the barrels is held well away from the face of the standing breech so it is only the hook at this stage that needs smoking. If the hook has been built up by spray fusion the visual effect and fit is the same, while a hook built up by welding will bear hard on that area, even if only a small section in the middle.

Obviously contact is made in the shiny areas where the smoke black is rubbed away between two rubbing or touching components. This then is the area to be relieved and a half-round needle file is one of the most universal tools for this job. The pattern of contact created by this sort of work tends to alternate, that is to say the exposed or shiny area gets larger every other time it has been worked on.

Eventually the barrels (less the extractor(s)) will start to contact the standing breech along their lower edge, and after smoking the contact marks should be evenly distributed on both sides to indicate that the barrels are truly in line with the action. As work filing the hook continues actually allowing it to move imperceptibly forwards, the ends of the barrels will move further down the action face until eventually the bolt starts to engage the bites in the lumps.

Things now start to happen quite quickly and it is easy to overdo the fitting of the hook even with such a dainty device as a needle file, so swapping to a small three-square scraper is a sensible option. By this time there should be contact right across the hook so the work now being done is to open it up to the required radius to fit the pin. On the barrels contact marks will be visible hard along their bottom edge, gradually fading towards the top as they are still sitting at a minute angle. The next step is when the locking bolt is almost fully engaged. Where an auto-safe is fitted it should operate fully, so the safety button is pushed right back by the action of closing the barrels, and have full forward travel to release the triggers. Then the top lever may still be a little more to the right of the centreline of the top strap than expected and the contact marks across the breech end of the barrels may still be hard at the bottom and light at the top. This is not at all a bad thing at this stage and it is best left as it is, the gun reassembled and a box of cartridges shot through it to settle everything down. The results of doing this are often surprising; the most obvious sign is the top lever nudging a little further towards the centreline, while a check with the smoker will confirm the fit. On no account should the gun be tested in this manner if the bolt is not engaged for most of its full depth.

If a little more work is needed, it is at this stage an almost imperceptible amount, sometimes no more than a polish with wet and dry on a round bar of a suitable size to fit inside the hook. If, with the breech end of the barrels soundly against the standing breech, the top lever is biased to the left, it is time to start work on the bolt (*see* page 91).

Some Problems with Rejointing

It is a certainty that, even with a gun where the looseness between barrels and action seems excessive, the actual amount of wear on the mating parts is surprisingly small. Movement, particularly at the end of the barrels, is exaggerated due to the leverage distance involved; this would be far worse if the lumps, or barrel sides in the case of an over-and-under, were not restrained by the sides of the action or action bar. It is a mistake to assume new cross pins or hinge discs need to be 0.020/0.030in (0.50/0.80mm) to take up the wear, as 0.004/0.010in (0.20/0.25mm) is more often the norm. With a gun using a cross pin, some idea of wear can be judged by measuring with a micrometer the captive and therefore unworn sections of the pin.

For another guide, with the barrels pushed down and forward on the action with the forend removed, measure with feeler gauges the gap at the breech. This approximates to half the amount of wear, not forgetting to add some on to allow for refitting. It is easy to be fooled; held against the light, a 0.004in (0.10mm) gap looks big enough almost to drive a train through.

After a gun has been rejointed on the hook/cross pin, and perhaps the bolt, the job is not finished. Quite often the forend iron has to be refitted against the back of the forend loop, especially if it had previously been 'tightened' to take up some early slack. Sometimes, new extractors will have been fitted with the barrels slightly off the face, and they will require refitting. After all that work has been done and the gun is clean and tidy, ready for the customer to pick up, the gunsmith really has earned his money.

CHAPTER 8

Lockwork

Introduction

To the user there is nothing more frustrating on a day's shoot, whether at game or clays, than a gun that misfires or refuses to fire. Somewhat more exciting, and something that is guaranteed to liven up a shooter's day, is a double discharge, that is, the second barrel going off on its own upon the discharge of the first. Guns, even those used for sport, are potentially lethal weapons and guns that have safeties that do not work properly and lockwork that is unreliable should never be used. Even allowing for all the care and safety discipline that normally applies on any type of shoot, someone could get hurt or even killed, because accidents can happen in the strangest of circumstances.

Lockwork problems are caused by either wear, maladjustment (usually the result of wear) and breakages. Very light trigger pulls, mistimed ejectors and intercepting sears that intercept when they should not are often the consequences of wear.

The most common breakages are strikers or firing pins and top lever vee springs. Occasionally there are more major faults such as broken ejector kickers, mainsprings and, very rarely, broken hammers or cracked bridles.

Reliability

In the reliability stakes some types of gun are better than others. Obviously, extra reliability is to be expected from a quality gun, but apart from these probably the most reliable type of shotgun is a late, good-quality British hammer gun with rebounding locks. It was extremely rare for a hammer gun to have ejectors, so any potential problems in that department are negated. The hammers always have a good length of throw and the strikers are sensibly sized and of simple design. Locks made by such masters as Stanton and Brazier are not just the

zenith of hammer-lock design but also beautiful pieces of work.

Unfortunately, hammer guns have had a bad press, much of it caused by worn-out, poor-quality guns and bad safety practices. All too often, hammer guns are now regarded as dangerous, even being banned from some shoots. The irony is that a quality hammer gun, especially one that can be opened and seen to be safe before letting the hammers down, is much better than many a modern gun with a simple non-automatic trigger lock safety.

The true sidelock shares many of the characteristics of the hammer gun and similar reliability is expected. Some have strikers that are either very small or somewhat long and slim, which do not stand up quite so well to hard use, but most problems are very minor. Side-by-side trigger-plate actions are rare and almost inevitably of good quality and so exhibit few problems apart from those associated with wear and age, which can affect any gun.

Although it does not carry a locksmith's name, this is a good-quality example of a hammer-gun rebounding lock. Note the elegance of the pierced (cut-out) bridle. Such attention to detail is rare nowadays.

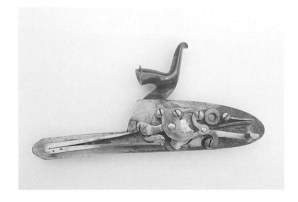

The side-by-side boxlock rates high in the reliability stakes, partly because there are so few parts to go wrong. There is, however, more chance of the mainspring breaking than with a sidelock as vee springs are at their best where they can be long and whippy relative to their width. Boxlock mainsprings, being shorter and cramped into a confined space, will occasionally fail, but, if they have been properly made, this will happen only after many years of faithful service and that can be a man's lifetime.

The over-and-under is the gunsmith's friend, partly because so many are not of really good quality yet get hard use on the clay pigeon circuit. Even some of the better-made ones have inherent faults that ensure reasonably regular trips to the gunsmith for minor repairs. It is true that rarely does anything dramatic go wrong, but broken firing pins and weak mainsprings are not at all uncommon with boxlock-type over-and-unders.

Other problems that sometimes bedevil all sorts of guns are sticking safeties, safeties that come on after the first shot and single-trigger mechanisms that do not select. A source of unique problems are ejectors that either do not eject or are mistimed. This is where the over-and-under scores as most were designed as ejector guns and problems are few, but some early side-by-sides have ejectors that, when well worn, are a bag of trouble.

Good-quality lockwork that has developed a minor fault is a pleasure to work on, but cheap locks can take a lot of time to get right. It is easy to finish up in the situation where repairs to a basic economy gun have cost in work time two or three times that of a far more expensive gun. Unfortunately, a customer with a 'Fred Bloggs Spanish Special Bargain Offer Model' expects low repair prices while the Purdey owner usually has a more realistic idea of the costs. Eventually you learn that some jobs are just not worth doing.

Over-and-Unders

Specific Problems

One of the most common complaints from an over-and-under owner is misfiring, which is due to worn or broken firing pins or weak mainsprings. Top-grade sidelock over-and-unders with vee mainsprings do not suffer with growing weakness in this component, but some do display firing pin problems; this is normally only if replacement pins have been fitted that are not exactly to the dimensions of the originals. The much more common

over-and-unders with helical mainsprings are a different matter as helical springs have a three-stage life cycle. From new, they weaken quickly then settle down to their working life. Eventually they weaken further and then start to collapse; the visual sign of this is buckling and a reduction in overall length. To make matters worse, as they buckle they then bind against the guide pin, adding a little more friction to both slow down and weaken the hammer strike. This process is accelerated by owners who do not use snap caps, and leave guns in store, sometimes for months, with the locks cocked and the mainsprings compressed. Obviously the answer is to replace the worn springs and there is no point in just doing one and not knowing the age of the other; always fit a matched pair.

The geometry of the lockwork on some over-and-unders is so badly designed that even under ideal conditions they only just fire a cartridge. Hammers are often of adequate length for a good strike but so often the bottom firing pin contact with the hammer is low down, weakening the blow. Also the pins are steeply angled and when a line of strike has to change direction it loses energy. Then the pins are long, adding weight and therefore inertia, of short travel, and sometimes fitted with fairly strong return springs. Add into this equation nothing more than a partly worn firing pin (as a striker is usually called on an over-and-under), and it becomes obvious why weak strikes against the cartridge primer are not uncommon.

A broken firing pin is a different matter – the broken nose of the pin usually falls out through the firing-pin hole in the breech face and the fault is all too apparent. Occasionally the pin will break and remain in place, so the fault is not obvious, but two-piece pins again absorb precious energy and cause misfires. The only answer, as with the mainsprings, is to ensure that the firing pins are in tiptop condition; whenever possible, it is best to fit original factory replacements.

Of the commonly encountered over-and-unders, two are worthy of special mention, for reasons concerning the firing mechanism. The obsolete Miroku 6/800 was fitted with vee springs and had hammers that struck directly against short firing pins situated in line with the chambers. Access to the firing pins was via discs set in the breech face. It was a design with great potential that at the time addressed many of the shortcomings of the popular over-and-under, and it was a pity it was not developed further. Its weakness related to the very short mainsprings, which would break without warning. The Browning Citori was a very similar

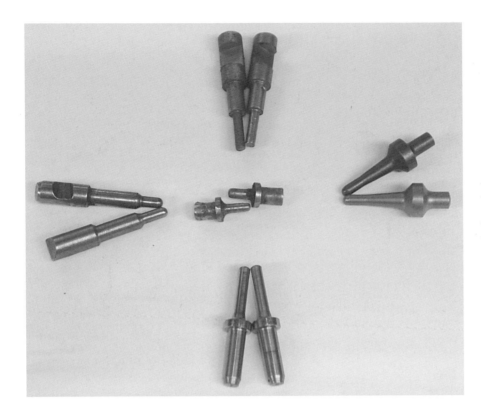

A selection of strikers/firing pins. On the right, Beretta's excellent double tapered pins.

gun, utilizing helical springs rather than short vee springs. Many shooters ultimately seemed to prefer a gun with a helical mainspring, which would have the occasional misfire, thereby giving an indication of things going wrong before complete failure.

The other make is Beretta, which has consistently ploughed a lone furrow. Its trigger-plate locks, for a moderately priced gun, are very neat, although some would view them as perhaps a little overcomplicated. One of the simple but really outstanding features of these guns is the design of the firing pins, which very rarely break. Although they are a little on the heavy side, they represent a lesson in good firing-pin design, with long, strong, tapered sections, well-radiused corners and steel that is superior to that used on many others.

The other common problem with a single-trigger gun is non-selection of the second barrel after firing the first. This will occur more often with mechanisms that rely on the inertia-block movement to select the second engagement. Some mechanisms will cycle mechanically even though they have an inertia block, and those have the edge on reliability. Usually there is very little wrong, and cleaning and lubrication may sort out the matter.

Where wear is found it is usually in the pivot hole of the connector arm supporting the inertia block. This can allow the block to slump down to a point where it is too low to engage the sear properly. The spring that pushes the inertia-block arm forwards may be broken or at fault. This is easily checked by pushing the safety forward to the off position and watching the block move forwards into engagement.

Most over-and-under ejector mechanisms are fairly foolproof, although the Winchester 101 type sometimes breaks ejector kickers. Such is the regard for this design that new redesigned heavy-duty kickers are now available. The Browning and Miroku break ejector extensions but not as often as the early Browning design, which has a 'D'-shaped engagement lug. (NB: *Spares, particularly for the Browning, are not cheap.*) The ejector action of so many over-and-unders is very positive, or, put another way, fairly sharp, which can eventually lead to the extractor breaking.

It seems difficult to believe at first but choice of cartridge can sometimes solve problems concerning both misfiring and trigger selection. Long gone are the days when the only cartridges readily avail-

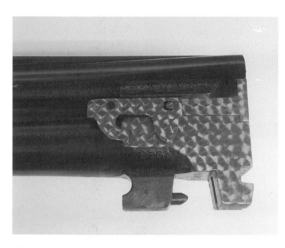

Very early Miroku that had broken the ejector extension. With spares no longer available, a Browning extension was modified to fit, requiring annealing, shaping, rehardening and refinishing. The final cost, of course, was more than fitting for a Browning.

able on the UK market were the Eley range, typified by the 'Grand Prix', 'Impax' and 'Maximum'. Search a bit further afield and there were the economy Sellier and Belliot, and also the 'Yellow Wizard' and 'Red Rival', at one time enthusiastically marketed by a firm called Young's of Misterton, which had a fascinating catalogue. The Eley were top standard and the others, while of variable quality, were at least dimensionally consistent. However, this is no longer the case. The quality of manufacture of some cartridges is appalling, with sunken primers and dished heads that reduce the impact and effective reach of the striker or firing pin. Hard primers are not unknown and trying a variety of cartridges in one chamber as a sort of makeshift gauge will show an interesting variation of fits.

In the days when black powder was king, the standard minimum load for a twelve-bore cartridge was 1⅛oz (32g) of shot. Later this became 1¹⁄₁₆oz (30g), but in recent years many shooters have moved to 1oz (28g) loads. Apart from the fact that they would actually be as well off using a sixteen-bore, the reduced recoil of some of these light-weight cartridges can be insufficient to cycle the inertia-block mechanism if there is the smallest problem in the locks. Even the physique of the user has an effect. People of substantial build may not find a problem, as they move little with the gun's recoil, which has the effect of operating the lock-work more firmly. Give the same gun and load to a slight youngster whose upper body moves with the

recoil and second barrel selection problems may rear their head. Use subsonic 'trainer' cartridges and, unless the gun has mechanical as well as recoil-induced selection, similar 'faults' can occur.

Springs

Making a Vee Spring Anyone can make a vee spring. The problem lies in making a spring that operates properly without distortion or stress and, as such, has a long and reliable life. The skill comes with experience and this is the reason why there used to be many dedicated spring makers, who would produce a gun spring of such quality that it would still be in use over 100 years later. Most will never achieve that standard of excellence but it is still worth a try.

There are short cuts, of course. There is no point in making life difficult and, if a gunmaker has standard springs available, they should be used. There are blank springs available that, after reshaping with a diamond file or a bench grinder, will cover a range of fittings. The 'set' of some of these can be a little hard and not as springy as it could be, but they are still a very good basis, and cut out a lot of work. Annealing them before filing to the required shape and fit, then re-hardening and tempering to get the springiness, improves them for proper operation and long life.

When no other option is available it means making a spring and the best springs are hand-forged so the shape of the spring follows the grain of the steel. Spring steel in untempered form is obtainable in a variety of widths and thicknesses, the most useful for most spring making are ½ × ⅛in (12 × 3mm) and ⅜ × ¹⁄₁₀in (10 × 2.5mm)

When making, for example, a top lever spring, most of the spring is forged and cut to shape while attached to the parent strip of metal – giving something to hold on to. Apart from a hammer, a hacksaw and files, the only other equipment needed is a propane torch, a steel block that serves as an anvil, small round-nosed pliers/mole grips and a peg cutter, which is easily made and fits in a brace and a former. Eye protection should always be used and leather rigger's gloves give good protection when dealing with yellow-hot pieces of steel.

The first move is to measure the broken spring from one end of the new steel strip and add on a small amount at either end so that the final length of the spring arms should be long enough to allow some fitting. Commonly, the top lever spring has a location peg so an allowance must be made for this. Because a spring is small and loses heat quickly it is best to have the propane torch fixed in the vice so

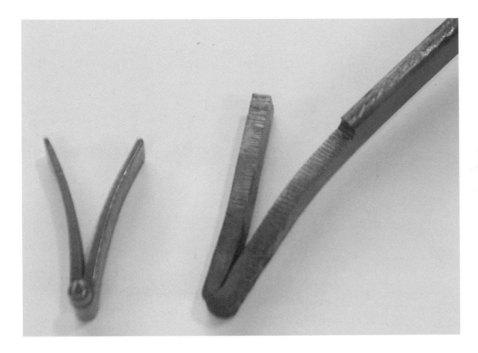

Partly forged top lever spring still attached to the parent metal as a useful handle. Alongside is a commercial spring, which is slightly oversize and requires some fitting.

you do not have to go through the constant ritual of putting it down and picking it up. This also means the torch can be kept running at working temperature with safety.

With a mark (file cut) to show the apex of the spring, the end of the spring strip is heated until it glows yellow. Using the small round-nosed pliers the strip can be bent back on itself; more than one heating may be necessary to achieve this. Then it is a matter of reheating and forging the bend so there is no longer a gap at the apex. Usually as you do this the arms will start to spread slightly.

After cooling, the now blackened steel can be marked with a scribe to outline the finished shape. Using file and hacksaw, the next step is to reduce the forging to the required width, leaving a roughly formed location peg. Next is to file the arms to taper form, working on the outside of the arms and, to get a good clean taper, finish off by draw filing with a flat needle file. The final result is a flattened vee spring, which may now be sawed free from the parent metal. With the spring still held in the vice the location peg can be trimmed to round with the appropriate peg cutter. For the final shaping, all that remains is to splay the arms to the correct form.

It is important that the apex of the vee remains tightly together without any gap so this is clamped up in the mole grips when the arms are heated.

They in turn are held in the vice and this then acts as much less of a heat sink than holding directly in the vice. When heated it is tempting to prise open the arms with a screwdriver but this will give an irregular curve of each arm, producing a stress point and, therefore, a weakness. For the initial vee setting it is far better to use a former, which can be made from any piece of suitable steel. This is then tapped down between the arms to produce a nicely curved vee.

Once it has been cooled again and is still in the soft condition, the spring can be trimmed with a needle file and polished with fine emery cloth. At this point, it is almost finished except for the slightly long arms, the precise length of which are determined at the final fitting after hardening and tempering.

Hardening and tempering is the part of this process that causes concern to most people, possibly because there is no precise measurement of temperature; it is simply a matter of eye, instinct and experience. However, the oil for hardening needs be nothing more special than a cheap-grade motor oil. A bean can half filled with this makes a suitable safe receptacle and it is easy to find the spring in what soon becomes blackened oil.

NB: *Do not use water for hardening or under any circumstances a glass container for oil. It is important that the oil is not too cold. Ideally it should be at around*

20 degrees centigrade or the ambient temperature of the workshop. For example, if the oil container has been left standing on a concrete floor overnight in the winter, it is best not to use it until the workshop has warmed up and the container has been off the floor for a while.

With the spring laid on a piece of fire brick (do not use ordinary house brick, especially if damp, as small pieces can shatter and fly in all directions), heat with the propane torch until it glows yellow. This is hotter than required, but from the time the brick is picked up to dropping the spring off it into the oil, it will change to cherry red. Tip it as near as possible into the middle of the oil, keeping your face well clear. The spring is cooled within seconds and when the oil ceases bubbling it can be fished out with long-nosed pliers.

The spring is now hard and brittle and, if it were put under compression at this stage, it would snap; this is why it has to be tempered. The best oil for tempering is whale oil, but this is no longer obtainable and it is, therefore, not possible to make comparative trials with mineral oils. Fortunately, modern spring steel is very forgiving – in other words, the variations between the precise oil used and the temperature it burns off will still produce a tempered spring. I have experimented with mineral oil (car engine oil), tallow and vegetable oil, and produced workable springs using all of these.

One clever amateur clockmaker I know puts his fine, hair-like coil springs in with the Sunday roast. He claims that not only does it temper them but they also come out a beautiful golden-bronze colour.

For the gunsmith, I do have reservations regarding the Sunday roast method. A shallow tin lid or a piece of sheet steel with a hollow beaten into it is quite good enough. The depth should be such that the spring can just be covered in oil. With the container on a suitable stand, spring and oil in place, it is heated up using the propane torch on a fairly low setting. The oil will boil and then burn. By playing the flame partly on the container and across the oil, but not directly on to the spring, the oil will gradually burn away. Obviously it is useful from the safety point of view to use the smallest quantity of oil necessary to cover the spring and the whole operation should be done in a well-ventilated area with a CO_2 fire extinguisher to hand. On a sunny day it is not a bad idea to do it outdoors.

Once the oil has burned away ('blazed off', in gunsmith's language), the spring should be left to cool fully and during this time should be protected from cold draughts. With a complex spring it is advisable to let the temperature drop quite slowly and this can be done by supporting the container on a fire brick when burning off the oil. The rea-

Blazing off a spring. It is important not to play the torch flame directly on the spring.

soning behind this is that, with a spring that has required a lot of work, it is better if it comes out too soft than too hard. If it is found to be too soft it can always be re-hardened and tempered, and the extra work put down to experience.

The heart-stopping part of the operation is now reached: testing the spring. This can be done with a spring cramp or, in the case of a simple vee top lever spring, by clamping it in the bench vice. If the testing is done in the vice, it is best to hold the vee end of the spring in a pair of heavy-duty pliers or mole grips to prevent it slipping and possibly flying out of the vice. The spring should compress evenly, then, when released, return to its original shape. With a top lever spring where the one arm is firmly anchored part-way along its length, it is acceptable, when testing in this way, for the other arm to do most of the work. Sometimes a small amount of setting will occur after initial compression and this is acceptable. If the spring arms deform from a vee shape, or even deform under compression, appearing almost to bend sideways, the spring is faulty and must be annealed and reshaped, or scrapped.

Assuming everything is acceptable, this leaves only the final fitting. The top lever spring can be offered up by compressing it in the vice and holding it with the top lever clamp. Where the arms are too long they can be trimmed and finished with a diamond file. Similarly, if the locating peg is longer than the mating hole in the underside of the top tang, this can be reduced to fit. Once the spring is installed and seen to be working correctly, all that is left is to take it back out, clean it up, hand polish it and refit it, then stand back and admire your handiwork.

A mainspring for an Anson & Deeley-type boxlock is little more than a scaled-up version of a top lever spring. Making a spring for a sidelock or hammer gun is much more complex, requiring not only the claw to be forged around a small piece of bar but very precise measurement between this and the location peg. If this measurement is either too long or too short, the lock might simply be unworkable. If the claw does engage with the swivel but is dimensionally incorrect the lock can freeze when only part cocked and even a lesser dimensional error might strain the swivel until it prematurely breaks. However, although it certainly takes time and sometimes more than a few failures to do one properly a well made vee main spring is a thing of beauty and a source of considerable satisfaction.

Repairing Vee Springs Normally this is something that is not attempted – certainly not for a gun in regular use. This does not mean that there are never times when a spring might be repaired. Take the cracked mainspring of a flintlock musket, for example. It is large and heavy, therefore fairly easy to work on and, most importantly to a collector, it is original. Cutting out the crack – and it is important to remove all the cracking – into a vee shape and welding with oxyacetylene is the best method. Filler metal can be thin strips of untempered spring strip or, if this is not available, $\frac{1}{16}$in (1.5mm) silver steel rod, although not quite as good, will suffice.

A similar method can be used if the spring is broken: clamp a supporting strip of steel on one side of the broken arm, then proceed as above. When the weld has cooled it is simply a matter of filing to shape, as the welding process will have annealed the spring in the area to be reworked. Once the original contour is reproduced it is best to anneal the whole spring at just above cherry red for about twenty minutes to ease any stresses. This can be done in a furnace or, just as easily, with a large nozzle in a propane torch and several fire bricks placed together.

After cooling, carry out the normal hardening and tempering. Although, as most of these repairs are for demonstration rather than regular use, it is best to anneal them a little on the soft side to give the spring the best chance of survival. To achieve this put into a container that leaves plenty of oil to burn off; always use it on a fire brick and even lightly play the torch across the spring as the last of the oil burns away. If it is near the end of the day and the workshop stove is on the wane, I would sometimes place the spring and fire brick on top of the stove so that it cools very slowly.

Then it is a matter of cleaning up, but not too well – the spring must look old and applying a little bit of artificial ageing might be in order. Experiment with acid treatments in small areas to get contrasting finishes, a bit like liver spots on the skin, observing all instructions in the use of these substances. Sometimes even artificial pitting in the form of a slightly blunt centre punch becomes necessary. Fine shot blasting can give a suitably matt finish as a basis for further work. And if it is a long-term project and real originality is required, bury the spring in a pot of horse dung and leave in the garden for a few weeks.

Coil or Helical Springs These springs, which operate under either compression or tension (the latter very rare in a gun), can be wound on a lathe. It is not easy to produce good helical mainsprings but for

Winding a spring on a lathe. Although I have never produced a spring as good as the manufacturers', it can be handy for an emergency repair.

lighter-weight springs, such as striker or firing-pin return springs and even recoil springs for self-loaders, it works well.

Apart from the lathe, spring steel wire of various diameters is required, plus winding mandrels and a simple tool through which to feed the wire on to the mandrel and to tension it as the spring is wound. The chosen mandrel should be slightly smaller than the internal diameter of the spring it is intended to reproduce. This is because the finished spring will expand slightly on the mandrel as the winding tension is released, and the amount of expansion is governed to a great extent by the tension originally applied. In other words, as so often is the case, it becomes a matter of experimentation and experience.

The mandrel is held by the plain end in the lathe chuck and the end with the anchorage hole is supported in a bronze ring held in a tailstock chuck. With the end of the spring wire inserted in the mandrel hole, a couple of turns of the chuck will ensure it is firmly held in place. Then, with the wire run through the guide, the clamp is tightened to provide tension. With the lathe set for a thread that matches the required pitch of the spring, the lathe is put into drive. To produce a really neat spring the first couple of turns should be formed parallel, prior to engagement of the lead screw, and when

the desired length of spring is formed the same is done at the other end. In order to get the hang of the method, it can be set up and the chuck turned by hand with the lathe power switched off.

To make a spring that operates under tension it is only a matter of choosing a thread pitch that equates to the wire diameter, so that the coils lie side by side.

Ejectors

Ejector springs will break, as will, very occasionally, ejector kickers, and then it is either a matter of replacement, if it is a factory gun, or making new parts. When there is a problem it is more common to find mistimed ejectors, that is, those that do not throw out the spent cases together. Sometimes both cartridge cases will be ejected clear of the barrels even when there is an obvious pause between the first and second ejection. If, instead, ejection occurs far too early, the edge of the cartridge case will strike the action face and bounce back into the chamber, giving the impression that the ejector has not worked at all.

Grit between the extractor legs or burred legs can cause dragging and weak ejection or the tendency for both extractors to operate with only one barrel fired. Then it is only a matter of cleaning and careful examination to detect the fault.

Nowadays the most common ejector system is the Southgate, deservedly so because of its inherent simplicity. The basic problems that can affect the Southgate type are in principle the same as those that will cause faulty ejection in most other systems.

There are a variety of ways in which the ejector mechanism is tripped on opening the gun after the fall of the hammer(s). Some have a cut-out to the side of the cocking dog or lifter, which engages with the ejector kicker (sometimes called ejector hammer). In others, rods or plates move forwards with the hammer to effect the same function. Wear on the contact points will result in mistimed or malfunctioning ejectors. When time is pressing – say, during the shooting season – those with thin plates that lie in the action slot of a boxlock alongside the hammers can be temporarily repaired by silver-soldering a piece of gauge plate into place with a long scarf joint so it is supported inside the action. Otherwise wear will need to be attended to by welding and re-case-hardening after shaping.

With the Southgate system, mistimed ejectors are most often caused at the contact point between the ejector kicker and the spring. The angle at the end of the spring controls the basic timing and this can be observed with the forend iron and ejector mechanism in place with the wood removed. It soon becomes obvious when the angled end of the

Plate-type ejector trip that lies alongside the hammer in some boxlocks. The scarf joint should extend back far enough so that the insert is supported inside the action.

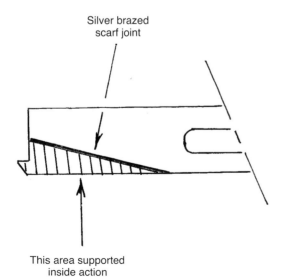

Silver brazed
scarf joint

This area supported
inside action

ejector spring has to be changed to produce clean, synchronized ejection. Altering the spring should not be done to overcome other faults such as wear.

Occasionally, odd things happen causing poor or intermittent ejection. One time the fault is not obvious is when the ejector kicker pivot pin is broken and one end is unsupported in the forend iron.

Trigger Pulls

The customer who comes into the workshop and announces, 'I shot forty-nine straight and missed the last bird. I think it's the trigger pull gone funny', is usually only fooling himself or subconsciously looking for an excuse. The usual reason for such a miss is not an alteration in the trigger pull, but the shooter heaving a mental sigh of relief that he has come to the last clay pigeon, rather than concentrating on it as he did on the first. That aside, trigger pull is an important factor in shooting consistently well; most rifle shooters are aware of this, but it escapes so many shotgun users, particularly, it seems, game shooters.

The weight of trigger pull is very much a personal thing, although anything below 3½lb (1.6kg) is, with a little wear, a recipe for a double discharge; many are set at 4½–5lb (2 to 2.6kg). A long dragging pull is to be avoided but with some cheap guns it is the only way to ensure safe operation. There is no better description of proper trigger pulls than the old adage 'it should go off [the trigger mechanism, not the gun] like breaking glass'. A translation of that could well be the trigger should snap off rather than pull off, even if the finger pressure required to release the trigger mechanism is referred to as trigger pull.

This trigger pull is a function of the depth and angle of the hammer or tumbler bent, the corresponding angle of the sear nose, the length of sear arm and the distance between the trigger blade/sear contact and its pivot point. Ideally the sear should slide out of engagement with just a little back angle to ensure that, if it is only partly pulled, then, when released, it will go back full depth into the bent. If this does not happen the next time it will be like a hair trigger and go off before expected. It will, however, reset to the full depth if the gun is opened.

When a trigger pull is very hard it may be the sear nose at the wrong angle so as it is drawn out of engagement it actually has to push back the hammer against mainspring pressure. Another source of hard pulls is point contact where the sear has been altered so its shape, when viewed from the side, is almost a point. This produces a high pressure loading (like a stiletto heel on a wooden floor), considerable friction and, therefore, a heavy pull. Replacement and stiff sear springs can also result in heavy or uneven trigger pulls. Very light – in real terms, dangerous – trigger pulls might be due to wear or the wrong angle on the sear nose, which means the hammer is actually trying to push it out of engagement. The initial movement imparted by the trigger is enough for it to then slide out of engagement the rest of the way on its own. This can also be caused, particularly with boxlock actions, by a poorly repaired stock where the distance between the top strap and trigger plate has been altered. If this dimension is smaller than original, the sear arm can be contacting the trigger plate and the sear nose already pushed partly out of engagement. If a stock has been weakened by the ingress of mineral oil, and the action pin and hand pin have then been retightened, compressing the stock, similar dimensional changes can occur. The signs that this has been done are the action pin slot being out of line and maybe the raised edge of the slotted head either filed down or fouling the top lever and the hand pin filed off to lie flush with the tang. When this happens, repairs should be carried out to the stock rather than giving in to temptation to alter the trigger mechanism to accommodate the stock fault.

With boxlock over-and-unders a common complaint from the keen competitive shooter is that the trigger pulls are uneven and seem to change when the barrel selection is altered. With many inertia-block systems the lock geometry does alter slightly between first and second barrel selection, but in real terms the difference at the trigger is not apparent to most users.

Assuming no wear or 'alterations' have taken place, one of the most common causes of different trigger pulls is the state of the mainsprings. With all locks the mainsprings power the hammers and the bent of the hammer or tumbler bears directly on the sear. If the force of the mainspring is altered, this alters the pressure on the sear and therefore the force required to pull it out of engagement. This is why it is important that, with guns using helical mainsprings, the springs should be replaced as a matched pair.

Uneven pulls may also be caused, although not very commonly, by the hammers where the bents have been cut individually rather than as a pair. This is normally a problem only on cheaper guns and can sometimes be corrected by aligning the hammers on a dummy pivot pin and cleaning them up on the surface grinder to make a matched pair.

With any form of alteration or repairs to trigger mechanisms it is vital to ensure that it is ultra-safe.

The relationship between the sear nose and hammer bent is the last-ditch defence against the gun accidentally discharging. It is better for a trigger pull to be a little too hard than too light, and slightly too long than too short. If there is an accident through possible negligence of the owner there is a good chance it will not be their fault.

Repairing Sears

If it is really worn or has previously been 'adjusted', the sear may be short, which has the effect of reducing the hammer throw and producing a condition where the overdraft, or degree the gun is opened to insert cartridges, becomes insufficient with the hammers cocked. When the gun is broken for cocking, the sears are heard to click into engagement, sometimes unevenly, but the barrels still move on their full travel then, when released, move back slightly. This is caused by the mainsprings pushing the hammers back for the bent to engage with the shortened sear. In an extreme example of this, it becomes necessary to hold the gun open against the spring tension in order to get the cartridge into the chamber.

When this happens the sear nose has to be built up with weld. Oxyacetylene welding using silver steel as a filler material is one option, but it should be noted that this takes considerable welding skill. The other, and safest, method is to make a new sear. Whatever method is chosen, after filing to shape, the nose of the sear should be hardened. Held a little way back with pliers that act as a heat stop, the end should be heated to cherry red and quenched in oil. Final fitting and setting the angle of the sear nose can be done in the same manner as resetting the trigger pulls.

Resetting Trigger Pulls

If worn, the sear nose is sometimes rounded and may be ribbed, but the line of the basic angle at which it was originally set can be deduced by study under a magnifying glass. To clean it up is only a stoning job – if a diamond file is used it will undoubtedly finish up too short – but this should not be by hand. Over the years, many strangely shaped sears have been produced that finish with limited points of contact and are generally unsafe. Having said that, my old mentor often used to reset the angle of a sear nose by drawing it by hand across a large oilstone of very fine grade and finishing by the same method with crocus paper laid on a flat steel plate. (Crocus paper seems no longer to be available but 2000-grade wet and dry is a good substitute.) He had fifty-odd years of experience, and he would never have dreamed of holding a sear in the vice and operating on it with a small square- or flat-section stone.

It is vital for trigger mechanisms to be reliable and safe, so it is worth being especially cautious when working on them. Inspired by watching a keen amateur cabinetmaker using a honing guide and angle jig for a chisel, I developed my own mechanical aid to achieve consistency. This 'sear jig' evolved from a sketch on the back of an envelope and was made to hold sears of varying length, right- or left-handed and with different-sized pivot holes. The slotted clamp allows the angle of the sear to be altered, to accommodate different lengths of sear nose and the necessary changes of angle.

A worn or incorrectly angled sear is set in the jig, the nose smoke blacked and the jig run across a flat plate such as a piece of gauge plate. This is to determine the basic angle for setting and to ensure the sear is set up square. Once this is done it is a matter of transferring the jig to a fine flat stone and drawing it down the length so the sear nose is shaped from the bottom up to the front edge. When a honed line appears across the sear nose, not necessarily covering the full contact area, it is best to smoke black it and try it in the lock to ensure it engages truly square with the bent. Assuming this is satisfactory, it is simply a matter of stoning until a clean angle is achieved, always finishing the job by smoke blacking and checking the sear contact is square and even.

To get a good indication of the finished trigger pull, the gun can be reassembled without the stock and the trigger pull tested with a small spring bal-

Reshaping a sear with my sear-honing jig inspired by a chisel-hone jig. Sometimes it is necessary to shim the sear arm if the pivot hole is out of line with the nose of the sear.

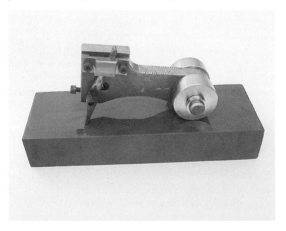

ance fitted with an extension. If the trigger pull is too hard the angle of the sear nose is too steep and needs to be reset; if it is too soft, it needs to be steepened.

The nose of some sears is a lot deeper than the depth of engagement with the bent and it is easy to assume that an unworn section is the original angle of the sear nose. This is not always the case, as it is not unusual to discover that they were made with a change of angle where they engage with the hammer bent.

Safeties

Gun safeties all have one thing in common: they should never be trusted. That is not to say that they may be of inadequate design or even fail to work properly, but it is poor safety practice to rely on any mechanical device to stop a gun being fired. The only truly safe options are to ensure the barrels are always pointed in a safe direction, or that the gun is unloaded.

There are essentially two forms of safety: by far the most common is the trigger lock, which locks just the trigger; the other, more expensive type is the one that will lock the mechanism in the event of a hammer slipping out of engagement.

It is generally assumed that the side safety button down by the trigger guard that is pushed to one side to prevent the trigger being pulled is in some way inferior to the safety operated from a button on the tang. This is not necessarily the case, as both may simply prevent the user pulling the trigger while the locks are actually cocked and ready to fire. In these circumstances a good-quality hammer gun with rebounding locks, where the tail of the mainspring pushes the hammer tumbler back into the half-cock position, is a superior arrangement. Contrary to common lore, such a gun with the hammers in the rested position will not go off if inadvertently snagged against a branch; the sear will simply re-engage the half-cock position as the hammer is released forwards. Only by having the trigger pulled, and hence the sear disengaged, can the hammer strike forwards to hit the striker. Still, even with best-quality locks I would not want to be in front of the barrels under such circumstances.

The safety system that is superior to these others is the type fitted with an intercepting sear, which is not often found on boxlocks but is fairly common on many sidelocks. Both the sear and the intercepting sears (or interceptor) have to be lifted to release the hammer and the timing of the interceptor is most important. The interceptor lifts first

An AYA sidelock with intercepting safety sear; one of the better safety devices.

so it clears the way for the hammer to fall when the sear is tripped. If the interceptor is late in lifting or the sear releases early, the fall of the hammer will be blocked. Clearly, in the event, say, of a sear being worn to the extent that it releases accidentally, the intercepting sear would block the fall of the hammer and thus prevent a discharge.

The intercepting sear is a safety system that can be highly recommended and it is a pity it is not more common, particularly on boxlocks. When faults occur it is mostly wear on the pivot that allows the intercepting sear to move forwards, so that it will accidentally foul the tumbler and prevent the hammer falling. Discarding the intercepting sear is not a 'repair' that can be applauded in any way, but it is surprising how many economically priced imported sidelocks are missing this part. Drilling out the worn hole with a tungsten carbide drill and sleeving it to obtain a good fit on the pin is an acceptable repair; often some work on the nose of the interceptor will be necessary to obtain the correct clearance.

The safety on inertia-block over-and-under mechanisms moves the inertia block and trigger connector backwards so it is held away from the sears. The effect is similar to a trigger lock system in that, while the trigger mechanism cannot reach the sear, the hammers are at full cock with only the sears holding them. This, in simple terms, is about as much contact as the end of a thumbnail. The better over-and-unders of this type do have a second or safety bent – what older hands would regard as a half-cock. Theoretically, should a sear be jarred out of engagement it will stop against the second bent, although to rely on this would be, perhaps, a little like playing Russian roulette.

The most common problem with over-and-

under safeties is breakage of the safety spring, sometimes referred to as the selector spring. The weakest of these designs are formed from spring wire with two flimsy arms. They are best, when possible, replaced with a spring made from flat spring strip. Another problem, not confined to over-and-unders but encompassing all guns of a similar safety-spring arrangement, is where the safety-spring pin does not locate in the correct place on the spring. If it does not ride over the shaped section of the spring to the proper position, it can easily, under recoil, slip back to the 'safe' position actually assisted by the spring.

Otherwise most safety problems are caused by lack of maintenance, especially lubrication, and sometimes by operator error – some shooters tend to wrap their thumb well forward over the stock and, under recoil from the first barrel, inadvertently push the safety button back.

Extractors

Over-and-under ejector guns are more inclined to break extractors than side-by-sides are, but then many over-and-under systems are almost brutally powerful. Where proprietary spares are available, some can be replaced with minimum fitting, while others will require the leg fitting to the dovetail, filing to fit the breech face and chamber radius, also cutting the recess for the cartridge rim. For this type of job a rim cutter is required and it is useful to have a chamber reamer to trim the final shape of the extractor radius to ensure alignment with the inside chamber wall.

Unless it has somehow been lost or accidentally broken, it is extremely rare to have to make an extractor for a non-ejector side-by-side gun. For an ejector gun, it is a different matter, as they do occasionally break. Usually, only one is broken, and this is most commonly the right-hand extractor, which gets the most work. For some imported guns, such as AYA, part-machined spares are available, but for an English gun it is a matter of starting from a raw forging. Welding or brazing is not an option for repair on an extractor used with an ejector mechanism. Even with careful choice of filler rod and post-weld heat treatment, they rarely last very long after this type of repair.

To produce the basic machined extractor from a forging the following steps are required:

• by holding the front extension of the forging in the lathe chuck and the end of the extractor leg centre drilled and supported on a running centre, the leg can be turned to the correct diameter;

- at the same time, the back of the extractor can be faced off leaving a small radius where the leg joins and an approximation of the radius turned at what will be the top of the extractor. Then the job is turned around in the chuck and the front face machined leaving it slightly thicker than the original for fitting;
- the next step is to mill the extractor to produce 'half a leg', once again leaving it slightly oversize to allow for the necessary hand fitting;
- the first piece to hand fit is the leg until the extractor butts up against the end of the barrels. Then, as with many bought-in items, it is a matter of hand fitting and rim cutting until it matches the depth of the rim in the barrels.

An extra complication can be the locating peg that prevents the extractor twisting out of line when at full travel. Visually lining up with an existing hole that is unseen with the extractor fitted is more than difficult – it is impossible. Applying an even layer of engineer's blue around the location hole and pushing the extractor hard against it will produce a mirror image of the hole's locations on the back of the extractor. Drilled and tapped 6BA is usually an adequate size for the peg. This can be secured by soldering on the front face so it enters the end threads; when cleaned up and hand polished, it should produce a tidy, flush-fitting appearance. The alternative is to have a projecting thread held with thread lock (commonly called Loctite) then peined over and filed back flush with the extractor face. The peening action fills in the half thread that would otherwise show as a slight gap.

Semi-Autos and Pump-Actions

To those used to the precision fit of a hand-finished gun, the investment cast parts and slack tolerances of semi-autos and pump-actions seem a recipe for excessive wear and subsequent mechanical unreliability. Despite the views of their detractors, however, they are usually quite reliable, and modern guns especially so. Simple wear is the greatest problem, rather than breakage. Realistically, by the time they are worn in many places until they have 'gone all chackley' (the dialectic word used by one of my contemporaries to describe the sound made by loose and worn mechanism), it is time to purchase something else.

Of those breakages that do occur, the most common is the cocking handle on semi-autos. As with extractor repairs, welding is an unreliable option, and replacement is by far the best method. The majority of cocking handles fit into the breech bolt but the Breda is one where it is formed as part of the bolt, so it becomes a considerably more expensive repair. The link that connects the breech bolt and action spring, when made up of two joined spring-steel strips, will sometimes split. These can often be brazed together quite successfully. The

Repairs to a locating peg. This half peg has been silver-soldered back into place. Others may be encountered formed as part of the main forging, but the AYA design does allow an economic repair.

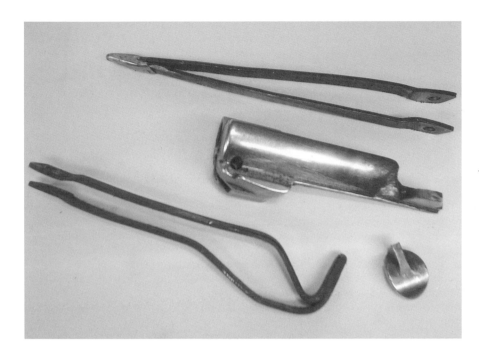

Broken cocking handle and damaged links; some of the more common failures of the semi-auto shotgun. With this one the cocking handle, being part of the breech bolt, is a much more expensive repair than normal.

latch that holds the bolt open sometimes breaks, but more often it wears out first.

Flat spring-steel mainsprings that operate not just the hammer, but also have fingers cut to operate ancillaries, sometimes snap off; this can happen with the one that operates the safety on the Browning Auto Five. They can be manufactured from thin spring-steel strip, then hardened and tempered. Occasionally firing pins break, and if a spare is not easily available, as long as both parts are found, it is easy enough to make another.

Misfeeds on gas-operated semi-autos are sometimes traced to the friction ring being fitted in the wrong position for the cartridge being used.

Usually most malfunctions are caused by a lack of maintenance. Gas-operated semi-autos require cleaning around the piston, particularly if they are to be laid up for some length of time, otherwise the accumulated residue from the gasses bled from the barrel can cause a misfeed for the second shot. Simply cleaning properly and lightly lubricating is the cure. Recoil-operated semi-autos will pick up grit around the magazine where the barrel lug recoils. How it gets there is always a mystery, but if enough accumulates it will add sufficient friction to slow the action, resulting in misfeeding.

At the other end of the maintenance spectrum is the owner who is fond of giving the workings a squirt of oil as a kind of maintenance cure-all,

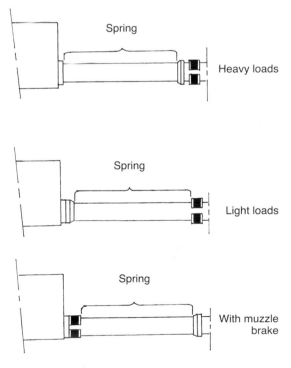

Settings for the friction ring on a Browning semi-auto. Misfeeds are often a consequence of these being set wrongly.

sometimes to the extent that it actually dribbles out of the action. With the ejection port a comparatively large opening in the side of the receiver, and the cutaway underneath for loading, there is ample opportunity for the ingress of foreign material. Heavily oiled workings will attract dust, grit and even pieces of dried undergrowth in a way that defies gravity. The sharp grit will chew away at the guide ways, while the other detritus will do its best to fill up the working clearances. Add a bitterly cold day, say, out pigeon shooting or on the salt marsh, and the dirtied oil will stiffen and becoming turgid and almost treacle-like. Malfunctioning is almost assured, especially with a semi-auto. Once again, a good clean is often the answer.

The pump-action, being manually operated, is not so sensitive and will keep churning out spent cases until the extractor(s) is too gummed up to grip properly or the action bar (or arm) breaks. This is more common with guns that have only one bar, which usually breaks where it joins the forend tube. Sometimes there is enough length to reshape the end of the existing bar(s) and still obtain an adequate length of movement on the forend assembly for full operation. This type of failure is not confined to the pump-action and also affects semi-autos. It is the one occasional fault with the otherwise sturdy and reliable Remington 1100, with metal fatigue being caused by extensive use.

Inevitably, tracing a fault will involve a full strip-down of the mechanism, and herein lies a problem. In 1988 a law was passed in the UK insisting that those semi-autos and pump-actions to remain classified as shotguns should have a fixed magazine

Broken action bar that operates both semi-autos and pump-actions. The single-sided bars are far more prone to breaking due to uneven loading than the twin bars.

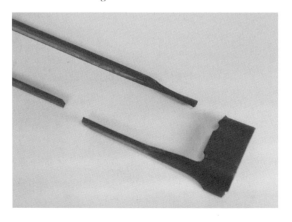

restricted to holding only two cartridges. The gunsmith's work was, by law, required to be checked by one of the Proof Houses. Guns such as the Mossberg Model 500 require the magazine to be disassembled from the receiver in order for the mechanism to be stripped out. Depending upon the method used to secure the magazine, a strip-down can become a very time-consuming job. It is also probably an offence to remove the magazine and later re-secure it without re-submission to the Proof House.

Another problem that sometimes occurs with three-shot conversions (two in the magazine, one in the chamber) is breakage of the magazine tube. This does not normally happen with steel tubes but some of those manufactured from aluminium alloy have failed due to the method of conversion. The common way to restrict a tubular magazine to two-shot capacity has been to impress an annular ring around the magazine at a defined point, so that it projected through the inner wall to restrict the travel of the magazine follower. With aluminium-alloy magazines it was best to anneal the area to be worked on, simply using a propane torch, but not everyone did this. If the tube was not annealed, or the annular ring was formed excessively deep, or in a vee form rather than comparatively wide and rounded, the stress at this point could weaken the magazine, until, with use, it failed. This becomes rather obvious when the forend and front half of the magazine and spring try to fly off. The broken stub of the magazine left in the receiver may, in the worst instance, have to be machined out to effect a repair. If the gun is to retain a two-cartridge-capacity magazine, any replacement magazine has to be converted or made to hold only two cartridges, and then needs to be re-submitted to a Proof House for certification.

As with inertia-operated over-and-unders, both gas-operated and especially recoil-operated semi-autos can be sensitive to choice of cartridge. Usually the heavy loads will work an action, but some light field or trainer load will not push the breech block back quite far enough, so it jams against the next cartridge on the carrier. The pragmatic answer is to use a cartridge of a make and loading that suit the gun.

The other related problem in this instance can be a dirty, rusted or rough chamber, which can produce enough friction against the case to cause a malfunction.

Lockwork problems on semi-autos and pump-actions are not simple, as, owing to the nature of operation, much more is involved than just the firing mechanism. Generally, where spares are no longer available and unless a repair can be effected fairly easily, the cost of the repair can easily exceed the second-hand value of the gun.

The Browning Auto Five is one gun that has something of a cult following, in spite of (or perhaps because of) its rather odd square-backed receiver. It is a very well-made gun, but the carrier latch is not available as a spare part. Ironically, most of the mechanism is substantially made, but this part is one of the few pressings and, as such, amazingly difficult, although not impossible, to duplicate.

The 1897 Winchester pump-action was made for fifty years until 1947, but it is not common in the UK and is actually becoming something of a collector/shooter's item. The shape of the receiver and the exposed hammer is quite distinctly Winchester and it can be found in solid-frame or take-down variants. It is noisy in operation, has a very exposed action, and is very solidly made and, as such, repairable if the customer is prepared to fund the cost.

At the time of the 1988 Act, many strange early actions surfaced in the UK. They not only required magazine conversion but also, often, proofing, having arrived in the country years ago, usually as a privately owned gun that, until offered for sale, would not require proof. Some were possibly quite rare – in relative terms – and interesting, while others were fairly awful. There are some guns that have definitely not improved with age and it is best to avoid any contact with them unless you have a desire to waste time and money. Among these early examples are the Winchester Model 11 and Model 40, the Marlin 1898 and the Remington Models 10 and 17.

The last words on the subject of lockwork must be 'safety first and last', even with the most seemingly innocent component. When something goes wrong and a gun is discharged by accident or neglect, the cry is either, 'I didn't know it was loaded', or, 'It went off when I didn't expect it to.' I was once sitting in the back of an enclosed Land Rover with other shooters and miscellaneous wet dogs when a gun was accidentally discharged next to me, fortunately up through the roof. It was an interesting, if deafening, experience and provided much food for thought.

CHAPTER 9

Minor Stock Repairs

Introduction

First, it is necessary to consider the different varieties of wood that may be presented for repair. The very best wood for gunstocks is walnut, and there can be few jobs more satisfying than repairing and refinishing a walnut stock and bringing it back to its former glory. Walnut is fairly light but strong, easily worked but sufficiently dense to hold good chequering and, in its more expensive forms, classically beautiful. It will endure years of handling with greasy, sweaty hands, stand some mineral oil leeching into the grain from the action and locks, and take dents, scratches and all that the weather can throw at it. And, when neglect has left it looking quite sad, it is often possible to restore it, for it is a most durable and forgiving wood.

What are the characteristics of the various types of wood used for gunstocks – particularly of walnut?

European Walnut (*Juglans regia*)

According to common lore, the very best stock wood is French walnut, actually wood from the European walnut tree grown in France. This tree, *Juglans regia*, is also reasonably common in the UK and across Europe and Asia, including Spain, Italy and Turkey. It has even emigrated to the west coast of the USA and other parts of the western world. The differences of texture and appearance and, to a certain extent, quality are caused by climate, soil content and rainfall. As a general rule, a tree that grows slowly and struggles a little against adverse conditions will produce better wood than a tree in ideal growing conditions on moist, rich soil.

There are regional differences in what is basically the same tree, but with a really good piece of wood it can be difficult, if not impossible, to identify the origin simply by appearance. A French or Turkish stock blank is usually light in colour with

dark veining, and a delight to work, particularly the French. Italian walnut tends to be light in colour with less contrast or veining, while English walnut is greyish in colour, again without particularly good contrast, and is one of the harder walnuts to work. English walnut is much derided – some gunsmiths declare that they would never have any in their shop – but a good piece can make a fine stock. There is also little doubt that English-grown trees have been taken to France for cutting up in the proper manner and returned to the UK with the appropriate French markings.

Types of Wood

American Black Walnut (*Juglans nigra*)
Like all walnut, American Black can vary from plain and a little soft to stunningly fancy. The Americans' favoured wood for some of the better grades is less restrained than European tastes, often containing fiddleback, burl or blaze – dramatic patterning where the wood is cut from the crotch of a tree. American Black does tend to have more open pores than European walnut, and this makes it more difficult when taking off a factory polyurethane finish, which will penetrate well into the wood. When oil finishing, these open pores will soak up oil like a sponge and, rather than giving a good contrast, can result in a smudgy appearance. When working this walnut it is easy to tell the difference as it smells different from the European.

Bird's Eye Maple
This is a scarce form of maple that was much favoured in the USA for really special one-off fancy rifles. It is most unusual in appearance, covered in tight whirling patterns, and its use in the UK is almost unknown, although British gunsmiths in the early nineteenth century were inclined to be more experimental.

I was once asked to repair a John Manton double-percussion muzzle loader stocked with bird's eye maple. The patterning was not confined to the stock but continued up through the hand and into the forend. With age it appeared to be trying to self-destruct, literally pulling itself apart from the inside – 'carroty', according to one particular cabinet-maker. Unfortunately, the stock had reached a stage at which its self-destruction would have been assured had it ever been fired.

Beech

Beech is used only on the very cheapest shotguns, although it was often used on military rifles as a substitute for walnut. (Perhaps it should have been mandatory for military use, given all those millions of rifles and all that walnut wasted, particularly in the First World War.) Beech is plain, strong and tough, very stable, light in colour and does not stain easily. It has a straight-grained flecked pattern and in appearance is particularly uninspiring. Simply from a point of view of strength, it would probably be better to use a good piece of beech than a poor piece of walnut, but the maker would have little chance of selling such a gun except at a rock-bottom price.

Cherry

A stocker of my acquaintance stocked two guns with wood from a wild-grown cherry tree and was convinced that this was not the first time it had been done. In weight it was similar to walnut, it was certainly hard and strong and it took very fine chequering. There was not a lot of contrast in the grain but it finished up an attractive dark, glowing honey colour.

Mahogany

This dark red timber would not normally be considered for gun stocks as it tends to be straight-grained and far too heavy. Mahogany stocks are sometimes seen on older guns re-imported from Africa, obviously replacements for the original walnut stocks; some of the workmanship on these guns is excellent.

Elm

English elm used to be described as 'poor man's oak'. It was once one of Britain's more common hardwoods, but mature elm trees have disappeared in recent years from the British landscape due to the ravages of Dutch Elm Disease. It was in the distant past used by gunmakers, but, unless a customer brings in something like a Cromwellian doglock musket, an elm-stocked gun is unlikely to be encountered.

Materials

Apart from the obvious tools – silica paper, wet and dry paper and several grades of wire wool – a few additional materials are needed to accomplish most stock repairs:

• french polish (sometimes called shellac varnish or button polish), in dark or 'garnet', clear or 'white';
• shellac or shellac sticks;
• Alkenet root;
• linseed oil (raw or boiled, it does not matter);
• walnut oil;
• white spirit or methylated spirits;
• terrabine driers;
• potassium permanganate crystals;
• walnut stain – not generally used for overall stock colouring, but nevertheless has uses for minor repair;
• stock oil – can be obtained commercially (Clive Lemon's products, among others, give excellent results) or made up by the gunsmith. One of my old mentors referred to his own brew as 'stock jollop': Alkenet root steeped in a mixture of walnut oil and linseed oil, with a dash of methylated spirits and some French polish to help it 'go off'. He never revealed the exact proportions of the mix so it is a matter of experimentation. It is easy enough, however, to make an oil for enhancing the grain and colour of a stock, even if it is not really suitable for a final finish: try 200ml walnut oil, 100ml linseed oil, 20mm French polish, left in a sealed container with Alkenet root measured by volume. A small plastic container cut off to give about 50-70mm by volume of root is sufficient. To a certain extent, the longer the infusion is left, the better. Shake occasionally and decant small amounts for bench use, leaving the root and the rest of the oil in the main container;
• stock finish – for use after an enhancing oil. French polish or button polish can be used, or a proprietary finish such as Birchwood Casey 'Tru-Oil'.

Minor Repairs to Stock Finish

Minor repairs include scratches, dents and abrasions, as well as a loss of finish, where, for example, a stock may have rubbed continually against something like a grit-impregnated waxed jacket. Scratches will vary in extent, from simple damage into the finish to penetration of the wood and torn grain. The nature of the original stock finish needs to be taken into account. An oil finish is repairable and infinitely superior to the spray-on varnish finishes; the very worst of these are the heavy polyurethane varnishes, which leave white marks where damage occurs.

Scratches in the Finish

For an oil-finished stock a scratch in the finish can be built up with French polish. Clean first with a spirit-dampened rag and, after the area on the stock has dried, apply French polish directly to the scratch using a pointed nib. This works something like a pen and it is therefore possible, with care, to fill or part-fill the scratch with very little overflowing on to the rest of the finish. The French polish will shrink during drying, so several applications to bring it up to the level of the original finish will usually be necessary. Between each application, rub down any surplus with 800 wet and dry on a small stick. When the scratch is filled, gently polish with '0000' grade wire wool. If the finished area appears dull in contrast to the rest of the stock, apply stock finish as necessary.

Sometimes it is useful to have a shellac varnish that has a thicker consistency than proprietary French polish. Traditional shellac varnish was made from one part shellac to seven parts alcohol but it can be made to the consistency of a filler by reducing the alcohol or methylated spirits to only two parts. It is best made in small quantities as it does not keep well and, like paint, settles out thicker at the bottom of the container, requiring stirring prior to use.

This shellac filler will fill small dents or scratches with only a couple of applications but is slow to dry properly and does still shrink. It is easy to be fooled into thinking that one application has filled a dent. A day after flatting down it may be found to have shrunk below the surrounding surface.

The other method of using shellac is to purchase cabinetmaker's shellac sticks, which come in a variety of colours and can be fairly well matched to the piece. On heating, shellac will become soft and, eventually, runny and the easiest way to melt it into a dent or crack is by using a small electrician's soldering iron, which is ideal for precision work. My experience is that the shellac 'filler' provides a much better bond, although it does take longer to do the repair.

Deep Scratches

When a scratch has torn into the stock, the wood has to be repaired before refinishing as above. If the damage is particularly deep, extensive or made up of multiple scratches, it may be necessary to rub down and refinish completely. Sometimes this can be done by working only up to the chequering line. However, where it is only minor or a single large scratch, the first requirement is to blend in the torn grain of the wood. This is done working over as small an area as possible; 180-grade silica paper wrapped around a rat-tail file makes a suitable tool. Once the torn grain is dressed out it will often be found that the wood is lighter in colour than the finished stock and has to be stained. For a small job such as this, walnut stain is appropriate and it is best to stain darker rather than lighter in colour as this leave a less obvious repair.

The repaired area can then be built up with French polish unless it is particularly deep. In this case, it may need some filling with shellac prior to the French polish.

As the repaired area will be larger than that for a simple scratch in the finish, the French polish needs to be applied with a cloth. It is not necessary over a small area to use a linseed oil-impregnated rubber, as a French polisher would. Instead, it is much better to use a small piece of lint-free cloth, folded to about the size of a small postage stamp, to apply the raw French polish directly to the repair. Once the repair is blended in, by alternate applications and rubbing down, the job is finished, as with a minor scratch. Sometimes even with minor repairs it becomes necessary to raise the grain to avoid a rough finish (*see* Chapter 10).

Abrasions

Abrasions require careful cleaning then rubbing down with 800 wet and dry, used wet, followed by building up with French polish or stock finish, whichever blends in best with the existing stock finish.

Damaged Varnish Finishes

Some so-called oil finishes are actually an application of varnish-based grain sealer followed by a coating or two of stock oil. They are characterized by a satin or even dull appearance and quite visible open grain. They can be repaired like a true oil finish; the only problem is that the repaired area may finish up with a contrasting better finish than the rest of the stock.

The thickness of other varnish finishes varies from a few microns to a lavish finish, such as that found on the Remington 1100. With the ultra-thin finishes, repairs using French polish are probably best; for the thicker finishes, applying spirit-based varnish is usually better, cutting back and building up the finish as appropriate. Where white marks are left by a cut or scratch in a polyurethane finish, little can be done except an attempt at disguise. In extreme situations a coloured spirit-based felt pen can be used to blend out the white marks, prior to building up the repair area with a heavy finish such as yacht varnish.

Dents

Dents may have simply pushed in the wood or, in more extreme cases, ruptured the grain, and the latter will not repair quite so neatly. Stock dents are raised by steaming – putting a damp cloth over the dent and applying a hot iron to the cloth over the dent. In many cases this will raise the grain almost to its original level. A domestic iron is somewhat clumsy for this rather fiddly job but, if nothing else is available, it will do. A soldering iron, either gas-heated or, more conveniently, electrically heated, is much easier to handle. Failing that, an old file ground flat and heated with the propane torch will suffice.

With many oil finishes, dents can be raised with the finish in place, although the best results are always achieved on bare wood. If the owner has been in the habit of wax-polishing his gunstock, a cloudy white appearance can result from the steaming. Fortunately, it is only on the polished surface and can be rubbed away and re-polished.

When the dent does not respond to steaming, usually because the stock has a really good 'best' oil or shellac finish, a few pricks into the dented area with a fine needle will help. The tiny holes can always be repaired with French polish or stock oil afterwards. With a dent in varnish, particularly some of the thicker finishes, steaming has no effect; it is just as well to fill the dent with varnish to bring it back up to the required level of the rest of the stock.

Broken Stocks and Forends

Degreasing

With an older gun, and sometimes a not-so-old gun, there may be oil in the head of the stock or the forend. With a clean, stripped stock this becomes particularly noticeable and requires further treatment before attempting any sort of refinishing or glued repair. Years ago, some gunsmiths used a very efficient industrial degreaser called

methylene chloride, but general use of this was banned and they switched to petrol. Modern petrol seemed somewhat oily (as well as dangerous) and did not do a particularly good job, so they turned to methylated spirits, which is alcohol and safer to use than either methylene chloride or petrol. The problem with oil is that it enters via the end grain and gradually seeps right into the wood. The only way to remove it is by the same process, using spirit to destroy the oil. It sounds a harsh process to impose on a good piece of wood, but if the stock is already really oily there is nothing to lose, and it works without any adverse effect if walnut-oil-based enhancing oil is fed back into the wood.

The head of the stock or forend can be immersed by standing in a clean bean can or similar, using the minimum amount of spirit to cover the affected area. This is always best done outside in a shady situation, and away from any fire, unless you want to have to make an embarrassing confession to the customer. After a couple of hours, remove the wood and leave it to dry in a sunny aspect if possible. Laying on a board or table under a warm sun does wonders, and if the oil content is excessive some of it will gradually leech out and can be wiped from the surface of the wood prior to re-immersing in the can of spirit.

Once the stock is a uniform colour, indicating that the mineral oil is removed, scrub it with warm soapy water, rinse and, after removing the surface water with a cloth, leave to dry in a warm environment.

If the weather is not suitable or it is necessary to carry out degreasing in a more controlled environment, the methylated spirits (some prefer

A whitened stock, in this case a particularly oily one, from a back-action hammer gun. The chalk was made into an emulsion with methylated spirits, then painted on. The discolouration is caused by the oil being drawn into the chalk.

methanol) can be mixed with powdered chalk to produce a paste that is painted on to the affected area of the stock. The spirit has the effect of neutralizing the oil, while the chalk – traditionally called whiting – will help draw oil out of the wood.

NB: *Any of these processes in the workshop should only ever be carried out in a well-ventilated area, away from heat or fire, and never near any other job that may be taking place, such as welding.*

Broken Toes

Guns not fitted with a butt plate or pad can finish up with a piece of wood broken off the toe as the grain is very short at this point. Where the wood has broken away – and assuming it is missing – the stock needs to be flattened off and this is most easily done on a disc or linisher. Shaping should follow the line of the grain and is achieved quite easily, as the original break will follow that line. The piece of new wood for the repair should be chosen to match the grain structure of the original, while colour on a small piece of repair wood is of little consequence as it can be colour-matched later after shaping.

The new wood should be flattened on the contact side so that the side view of the grain lines up with that on the stock. At the butt end of a stock there is rarely any contamination such as oil so it is possible to use a waterproof powdered resin glue, which, with a good contact area, does not leave a glue line.

Clamping the repair wood in place is not easy, as a stock is both curved and tapered. A flat piece of wood with a raised edge screwed on, which can be held in a vice, makes a good base. Another strip of wood about 1 × 1½in (25 × 38mm) can be clamped

as a wedge to hold the repair wood in place. Sometimes it is necessary to use clamps in other positions to hold the job securely, always making sure the jaw is padded to avoid damaging the stock. It is best to use too many clamps to hold everything in position than too few. A useful, flexible third hand is also easily made by cutting strips of rubber from a car or motorcycle inner tube to bind around the job; this will sometimes suffice without any other form of clamping.

If possible, leave gluing jobs undisturbed until the next day to be quite sure that the joint is really secure. A repair to the toe of a stock can be rough-shaped on a linisher or disc then finish-shaped with a fine rasp before raising the grain and final finishing. Colour-matching is not too difficult and can be a combination treatment of potassium permanganate solution, stain and enhancing oil. Done carefully, the repair should almost be invisible, which is achieved more easily on a dark stock than on a lighter-coloured piece of wood.

Cracked Hands

Occasionally a crack along the grain will be found running from the head of the stock in a line through the breech pin and the long hand pin. If the gun has been used for a long time in this condition, then the breech pin hole will have become slightly oval as the stock has tried to spread apart under recoil. Simply gluing up the stock will not be sufficient in this case because the integrity of the fit of the pin is vital in holding the stock firmly to the action.

The head of the stock usually needs degreasing because even a trace of oil will prevent the glue bonding with the wood. Even when degreased,

Stock clamped to a flat piece of wood with a raised edge to hold the newly inserted repair in place while gluing. You can never have enough clamps.

This stock had split along its centreline, following the run of the grain. Note the excess glue to be cut away.

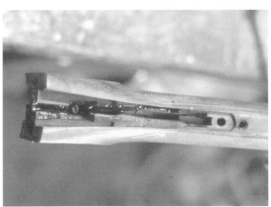

because this part of the stock has nearly always suffered some contamination, it is better to use a two-pack epoxy glue. Even though it leaves a visible glue line, with internal cracks this is of no consequence as the repair will be well hidden. After degreasing, the existing crack should be pushed apart and glue run in for the full length. To facilitate this, warm the glue first – prior to mixing the two parts – either on a radiator or a sunny windowsill. Once plenty of glue is in the joint, it can be clamped up to dry. Some people remove the excess glue at this point, but it can be left to squeeze out all along the joint line and, if this is not excessive, left to dry, then cleaned off with a small chisel.

The next step is to repair the breech pin hole. First, the old hole has to be cut out to fit in sound wood so a new hole can then be drilled to fit the pin. A drill with a pilot that fits the existing hole is the best way of achieving this, and the major diameter of the drill should just fit within the square cut to accommodate the trigger-plate turret. Once this hole is cut, a slightly overlong dowel can be made to fill it. Walnut does not make a good dowel for this type of repair as, in the finished form of a hollow cylinder, it is inherently weak. Beech makes a fairly good dowel, but much better is a piece of really hard, stringy willow such as cricket bat willow. Once the dowel is turned, a sharp spiral cut along its length will give extra grip for the glue to bond.

With the dowel glued in place and protruding slightly at either end, the action can be refitted by smoke blacking the strap and carving away the dowel with a really sharp chisel until good contact is made. The same applies to refitting the trigger-plate box. Then that leaves drilling for the breech pin. The breech pin is invariably tapered, so it is a matter of choosing a drill bit that is a good fit for the thread and minimum diameter of the pin. It is then necessary to hold the action in place with the stock clamp so that the action is really firm against the head of the stock, just as it is intended to be when in use. To mark out the top position of the breech pin hole, turn a soft-steel centre punch that just fits the hole in the strap and can be canted slightly out of line so that the indent can be made about 0.010in (0.28mm) from the true centre position towards the action. When the hole is drilled, this will ensure it is pulled hard against the back of the action. The hole above the trigger-plate box also requires an indent made with the centre punch, and the position of this can be detected by smoke blacking the box and clamping it into posi-

tion. If smoke blacking does not give a clear enough indication of the hole position, borrow some lipstick (the darker the better) and use that as a substitute for smoke blacking.

Drilling this hole has to be done on a drilling machine with provision for clamping a machine vice to the table. A pointed brass bar is clamped, point upwards, in the vice and lined up with the drill in the chuck. The bottom centre punch indent is then located on the pointed bar and the top indent lined up with the drill bit. A second deep-jawed vice can be mounted on the side of the drill table to secure the stock in position, then it is simply a matter of drilling through the dowel until the drill bit just touches the top of the brass bar. It is essential to stop at this point, otherwise the drill bit will try and slide down the taper of the bar, with the possibility of breaking the bit and the certainty of spoiling all the repair work done to date.

The hole is then finished with a hand brace and, should the long hand-pin hole need repair, the same method can be used but without offsetting the hole and after the action pin has been refitted.

Split Forends

Side-by-side forends may be encountered with wood split from the sides and, very rarely, with a longitudinal split along the middle, although this type of damage is not so uncommon on over-and-under shotguns. The other place over-and-under forends will be found with cracked or missing

Glued repairs should have as big a contact area as possible. Cutting back at a shallow angle towards the inside of a stock or forend will achieve this and give a better joint.

New wood

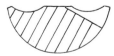

Section through forend showing
long shallow angle method of repair

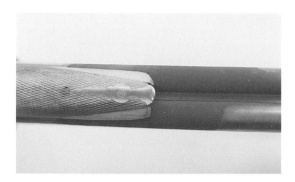

A repaired forend partly shaped on the outside and being refitted to form around the barrels. Note the smoke blacking on the barrels to mark the contact points. This can also be done with engineer's blue.

wood is at the back top edge in designs where the ejector kicker sits just inside the woodwork. These repairs are similar in that they involve thin sections of wood almost inevitably split along the grain.

The trick for most of these repairs is to increase the contact area of the glue line so that it is greater than the original break. This is done by cutting at an angle from the original break for the replacement wood to fit. With a forend where the break runs into the chequering this is a considerable advantage in effecting a near-invisible repair. If the angled joint for the new wood follows a line so that, when finished, this joint line will be at the bottom of the chequering cut, it will be almost undetectable, particularly if the chequering runs almost to the end of the forend.

With wood broken away at the rear of an over-and-under forend, a similar method of repair can be used, except there is always going to be a thin line just visible. This can be minimized by careful matching of the grain and the use of powered resin glue.

The biggest problem that can occur with forends is a longitudinal split almost down the middle, where the two halves not only require gluing along the line of the split, but also tying together to impart greater strength. With a side-by-side forend this is not such a problem, as there is a reasonable thickness of wood to work comparable to the overall dimensions. With over-and-unders, however, this can be one of the thinnest sections. Where possible, a tie – a piece of wood let in at right-angles to the break – should be fitted on the inside. If the forend is completely split then it is best to use two ties. If the only option is to fit a tie from the outside it can be dovetailed in where a chequering panel might be used to disguise its presence. However, with the grain at right-angles, even carefully colour-matched, the repair will be visible from some angles.

Stock Extensions

Introduction
A stock can be extended by fitting a butt pad where previously there was none, or perhaps to replace a thin butt plate. When the extension exceeds 1in (25mm), and particularly if it is a quality gun, it is far better to extend the stock with wood, as even a butt pad in excess of ⅝in (15mm) can look out of place on a shotgun. Vulcanite – commonly substituted for Ebonite – makes a smart thicker butt plate but, as a rule, on most stocks, it should not exceed

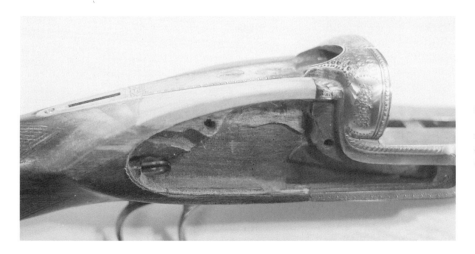

This stock had a piece broken out above the lock. The new wood, which is visible, was grafted in as an angled repair after degreasing.

½in (13mm). Thick lumps of Vulcanite of up to an inch or more do absolutely nothing for the appearance of the gun and can unbalance it, as Vulcanite is heavier than the equivalent walnut. Much of the success of a job is a matter of proportion and styling, and it is far better to do a visually appealing job than simply graft on a lump of rubber for convenience.

The very best material for a butt plate is black buffalo horn, which, in spite of its name, has some interesting and subtle variations in colour. Being a 'living' material, like wood, it is very complementary when mated up to a good piece of walnut. Probably the worst example in terms of appearance, no matter how well executed, is a ventilated butt pad with white line spacers fitted to a good side-by-side – a bit like equipping a classic sports car with white-wall tyres! In short, any extension to a stock, if done properly, should not advertise its presence.

Butt Pads

The most useful butt pads to keep in stock are dark red or brown, plain or leather effect, in sizes ⅜in (10mm) to ⅝in (15mm). Pads made of Sorbethane,

When a butt pad is to be refitted it is rare to find the screw holes lining up with the existing ones. This can be worse when butt plates with raised dowels were a fitment. One answer is to plug with simple dowels and glue in place with fibreglass resin, using extra resin to fill the dowel cut outs, then flattening off.

the material used in the soles of many training shoes, are very efficient at reducing felt recoil, but it is difficult to get a really good finish where they align with the stock; with the early ones, it was a little like trying to nail a jelly to the ceiling. They are good for applications such as heavy-recoiling rifles, but, no matter how efficient, a butt pad should not be used as a substitute remedy for a poorly fitting stock.

Butt pads are made to be fitted flat, so the butt may require flattening, which can be done on a sander disc. The line to finish to can be defined by marking with masking tape, which is then left in place to protect the stock until virtually all the fitting is complete. Surplus screw holes in the stock should be filled with tapered walnut dowels, which can simply be carved with a chisel and glued into place, the surplus dowel cut off afterwards or done on the sander disc as part of the initial preparation. This blanking of holes that will not be used is not so much for the customer's benefit, but for the next gunsmith who, at some time in the future, may work on this part of the gun. It is disappointing to take off a butt plate to find multiple holes, or holes reinforced with glue and broken matchsticks. Doing the job properly, even though it may remain unseen, perhaps until the gun is scrapped, is a matter of pride.

With the butt prepared, the pad can be offered up and the fit on the face checked by holding it up against the light. If there are gaps between the pad and butt, it is necessary to clean up the base of the pad on the sander disc or linisher to achieve a good fit. The fit can be checked by applying engineer's blue to the pad, holding it against the end of the stock and moving just a little side to side. High spots will be shiny blue, gaps natural wood. Once a good fit between the two mating faces is achieved, the screw hole alignment is found by holding the pad on the stock in the required position and marking out one hole position; for convenience, this is usually the hole at the heel. It is important to pre-drill the hole using a drill bit the size of the core of the screw and to the full depth the screw will engage the stock. Lubricate the screw thread with beeswax and the head of the screw, where it will push through the butt pad, with washing-up liquid. A little of this kept in a pot will eventually thicken to a slimy jelly, which works very well to ease the head of the screw into the pad (and, of course, is easily washed off). To avoid 'chewing up' the pad holes, use a round-shank screwdriver.

With the butt pad secured by the heel screw, the toe screw position can then be marked out by

This double muzzle loader had been drastically shortened for a youngster. Note the effort made to match the characteristics of the grain and the fine joint line achieved by using a resin glue.

pushing a scriber through the hole in the pad. With the pad released slightly and swung to one side, this hole can then be drilled to accept the screw. As a screw tends to push up the wood around a hole, each of these holes should be chamfered with a forty-five-degree chamfering tool to prevent this affecting the fit.

The bulk of the surplus projecting pad is most easily removed on a sander disc or linisher. To finish, a double-sided rasp, coarse and fine, is one of the best tools and should be held at a slight back angle, as the rubber will tend to spring outwards. This back angle means the pad can be cut right up to the edge of the masking tape protecting the stock. With a ready-finished stock it is best to stop at this stage and just take the sharp edge off the pad where it comes up to the butt. This visually has the effect of blending it into the joint rather than appearing as a tiny step.

If the stock is to be refinished, even only as far as the chequering line, then the pad can be blended in to the line of the stock using silica paper on a block.

Wood Stock Extensions

The use of wood can be regarded as the classic method of stock extension, and is much more satisfying than working with a rubber butt pad. Although many of the same principles apply, and a joint line will be visible, if it is done properly the line will be neither obvious nor visually offensive. Also, if care is taken to match the grade and grain flow of the wood, rather than carrying out 'just a stock extension', it can even become a subdued but neat piece of work worthy of admiration in its own right.

As with fitting a butt pad, the end of the stock needs to be flattened and this is achieved by using the sander disc; the mating face of the oversize, rough-shaped extension is also sanded flat. For guns without a stock bolt, the extension is normally left solid and drilled to accept two countersunk screws so that the screw heads will be at least $\frac{1}{4}$in (6.5mm) below the finished butt end. It is worth noting that occasionally, as an expression of the stocker's or gunsmith's skills, the end of the stock will be curved to accept the extension, the curve matching the finished curve of the butt.

The stock should always be pre-drilled using a drill bit the size of the core of the screw. Using the extension as a drill jig, drill the hole at the heel end of the stock, hold in place with the appropriate screw, then drill the hole at the toe end of the stock, ensuring the holes are along the centreline of the butt. Dismantle and chamfer the holes on both of the mating parts so that there is no wood pulled proud by the screws to spoil the joint. Then, using a waterproof powdered resin glue, the two parts can be screwed together; with glue on both mating faces, it is useful first to rub them together, thereby ensuring complete, even coverage, and pushing out some of the excess glue prior to screwing up tightly.

Dowels will be needed to cover the screw heads, and these should be turned on a lathe from matching walnut. This does not require the use of a specialist wood lathe; it is quite practicable to turn rough-finished wood parts on a metalworking lathe in the same manner as a piece of brass. With dowels the correct size to fit the holes over the screw heads, cut a shallow slot up the side of each one with a hacksaw blade. This is essential when fitting a glued dowel into a blind hole, to let the air escape from in front of the dowel as it is pushed or tapped into place.

Twenty-four hours later, once the glue has set, the extension can be shaped quite close to the finished dimensions, using a spokeshave and finishing with a rasp and then silica paper on a block. It is then best left for a few days, as the moisture content of the butt and extension may not be the same. Doing this avoids a step appearing later between the two parts. A half-round rasp is particularly useful for shaping the end of the butt. Some of this work may be done with a spokeshave, but it needs to be a super-sharp blade to cut cleanly across end grain. Prior to raising the grain, colouring and finishing at least up to the chequering line on the hand, some kind of grip will need to be cut in the butt face. If it is a quality gun then it is worth chequering with, for practical purposes, a fairly broad cut, say fourteen to eighteen lines to the inch. If a more economic job is required, a metal chequering file can be used to achieve a simple but pleasing pattern by cutting straight across the butt, leaving uncut contrasting areas at toe and heel.

For a gun with a stock bolt, which includes most over-and-unders, a through hole will have to be cut, and this looks tidier if made to match the existing butt cutaway. This can be done after the extension is glued in place by drilling through with a spade-end drill bit of about 1in (25mm) diameter, then using a rasp file to profile the inner edge to match the existing stock hole. The only other difference is the way in which the extension is held to the main butt. If the extension is short it is best held on with the screws that hold the butt plate in place, using that same plate as a drill jig for the screw alignment. Screws slightly longer than the originals may be required.

When the extension is comparatively long, say, over an inch, it requires separate screws, which can be accommodated by offsetting them to the butt-plate screws. I have, however, known it possible to use two sets of screws in line with the extension screws countersunk very deep and these holes dowelled for the butt-plate screws. If a stock with a large cutaway has been cut back particularly short, leaving little wall thickness of wood to work with, this is sometimes the only method possible.

CHAPTER 10

Major Stock Repairs

Stock Refinishing

Stripping the Woodwork

This does not necessarily qualify as a minor repair, but when there is a large amount of minor damage it may become the only practical, or best option. At other times it can be at the customer's request, particularly where there is a good-looking piece of varnished wood, which will look even better with an oil finish.

To achieve a refinish, all damage and original finish must be removed. Assuming the damage has been rectified, necessitating a complete refinish, both oil-type and varnish finishes can be taken off with a paint stripper such as Nitromors. (However, take careful note of the maker's safe handling instructions and always wear rubber gloves and eye protection.) It is easiest to do one side of a stock at a time and the same applies with a large forend such as that of an over-and-under. With a heavy varnish finish, scraping into the surface of the finish with a cabinetmaker's scraper

or even the edge of a knife blade will speed up the process.

After a liberal application of paint stripper (including the chequering) has lifted off some of the finish, douse under warm water, scrub any loose residue away with a small, stiff-bristled nylon brush, leave to part-dry, then reapply the paint stripper. This amount of work will vary considerably depending upon the original type of finish, but eventually it will finish up as completely clean wood, including in the chequering, which is always something of a revelation. The woodwork may not have been seen in this natural condition for decades. Finish off by scrubbing with warm, soapy water, rinse clean, then wipe off all surplus water and leave to stand in a warm, dry place for a day.

Preparation for Finishing

By this stage, with all the work so far completed, it is easy to assume that is nothing more to do now but get some finish on to the wood and watch the

Two views of a stock, one showing the stock as received, grubby and oil-soaked (left), the second showing what can be done by de-oiling and cleaning off with paint stripper (right). Walnut is a wonderfully durable wood.

beauty of the grain start to appear. However, there is one important part of the preparation still to come: raising the grain. As both stock and forend are not flat pieces of wood, but complex shapes of varying thickness in places, this shaping cuts across the end of the wood grain. Under certain conditions of dampness the end grain will raise up, producing a rough surface, which is very undesirable. Usually after treatment with paint stripper and alternate flushing and drying, if the grain is going to stand up it will already have done so and can be flatted down with 400-grade wet and dry and water. If new wood has been inserted into a stock as part of a repair, it is a certainty that the end grain will be very visible.

One way of encouraging the end grain to stand is to wet the wood and gently play a propane torch set on a soft flame around it, but not directly in contact with the wood. The aim is to produce a fast drying process, and if any of the end grain is likely to stand up, this will induce it to do so; it can then be flatted down.

Colouring Stockwood

Sometimes, after all the preparation is completed, it becomes obvious that the wood is rather lighter in colour than is normally desirable for a gunstock. In the UK, certainly, stocks and forends are preferred in the darker walnut colours. With some varnish finishes, all the colour is in the varnish; when it has been cleaned off the stockwood is as pale as sapwood and requires quite an amount of work to make it presentable. In this case, there is no option but to enhance the colour of the wood.

Staining with either spirit-based or water-based stain is best only on small repairs. For large areas, such as complete stocks and forends, it is best to change the colour of the wood using potassium permanganate. A solution of this, made up from half a teaspoon to 100ml of water, produces a bright pinkish-purple liquid, which looks quite inappropriate. However, once it is applied with a small piece of cloth to a stock or forend, the wood changes within seconds to a rich reddish-brown. (It will also stain fingers to give the impression of a die-hard nicotine addict.) The other side effect of using potassium permanganate is that, even after all the careful preparation, it will sometimes raise the grain again!

It is best with this treatment to apply thin coats, rubbing down very lightly after each application has dried, rather than making up a thick mixture and trying to do the job in one go. It is difficult to get an idea of how much the wood should be darkened to achieve the desired final colouration, particularly as each piece of wood will react differently; experience help here. As a rough guide, the potassium permanganate treatment will become significantly lighter as the stock dries, and if in doubt you should err on the dark side for best results.

Oil Finishing

Some wood – particularly the best – 'contains no artificial colouring', but whether stockwood is treated or not has no effect on the oil treatment.

Gun enthusiasts tend to be great believers in 'palming in' boiled linseed oil as the only true and reliable method of finishing a stock. This procedure involves rubbing in small amounts of linseed oil and working it into the wood with the palm of the hand until it feels warm. Unfortunately, it can easily take up to three months, depending upon the time of year and the nature of the piece of stockwood. Also, linseed oil on its own is not particularly durable, takes a long time to dry to a hard finish, and does not bring out the best in the grain.

One way of achieving a good finish in a reasonable amount of time is first to apply enhancing oil. One application each day for four or five days will bring out the colour and contrast in the grain. The first application should also cover the chequering, and can be applied with a toothbrush, with the surplus being rubbed into the rest of the stock with a small piece of cloth folded to the size of a postage stamp. Subsequent contact with the chequering should be avoided, otherwise it will clog with oil and look messy. After the first application, there is a choice: either continue using the enhancing oil with a little terrebine driers added, gradually building it up until a smooth overall finish is apparent; or switch to a proprietary oil-based finish such as Tru-oil once the desired colour and contrast have been achieved, in which case a good finish can be produced in about a week. Using the enhancing oil with added driers takes about three weeks, depending once again on the piece of wood and the ambient temperature. Some areas of a stock will absorb oil at different rates to other parts, dependent upon the grain.

Oil should not be allowed to build up on the surface and, where it does, it should be removed by rubbing down with 800-grade wet and dry. Once an even coating has been achieved, further applications can be made by rubbing in very thin coatings by hand ('palming'), or, for a quicker, hard final finish, applying French polish with a rubber (a small piece of folded lint-free cloth dampened with linseed oil).

Everyone has their favourite methods and, to a certain extent, the way a stock is finished will depend upon the quality of the wood or gun, and the price the customer is prepared to pay. After all, regardless of good intentions, there is no point in trying for a Purdey finish, say, on a cheap imported gun with a plain part-sapwood stock. It does not make economic sense, neither would it be fair to the customer.

Varnish Finishing

For some factory guns, the effect of a final soft satin finish can be achieved by rubbing the stock with a slightly abrasive brass polish. If a more vigorous treatment is required, chrome polish or cutting compound used to restore car paintwork can be used. To finish, traditional rottenstone powder mixed to a fine paste with linseed oil thinned with a little methylated spirits and rubbed well on to the stock, then polished off, gives the expected results. Revarnishing a stock brushed by hand is not easy, but a good cabinetmaker who sprays furniture will usually have a matt, silk-type finish that is very acceptable. This can usually be applied after colouring a stock or after using enhancing oil, providing the oil has had time to dry. A very thin first coat can even be sprayed across the chequering, but after that the chequering should be masked off the avoid clogging. Usually two to four coats are needed depending upon the openness of the grain. My local cabinetmaker uses a two–part acid catalyst lacquer, which gives a pleasing, durable finish superior to many factory gun finishes.

Grain Filling

With an oil finish on a fancy stock where the grain lies in different directions, oil will be absorbed at different rates. American Black walnut, with its tendency to a coarser, open grain, can absorb stock oil like a sponge. Sometimes, therefore, it becomes necessary to use grain filler – not the spirit-based, varnish type obtained from the DIY store, but an inert powder that, mixed with a little stock oil, can be worked into the open grain. There is no proprietary powder used specifically for this purpose and many gunsmiths' favourite filler is talcum powder! Alternatively, if a darker filler is required, rottenstone powder is equally suitable. However, I prefer, wherever practical, to keep working with oil and a little hardener.

Chequering

Introduction

Chequering is the final finishing touch that elevates a wooden stock of subtle shape, colouring and elegance to a thing of striking beauty, yet the origins of chequering are purely functional. It was intended to provide grip – a useful non-slip surface – but gunmakers over the years turned this practical feature into an art form of its own.

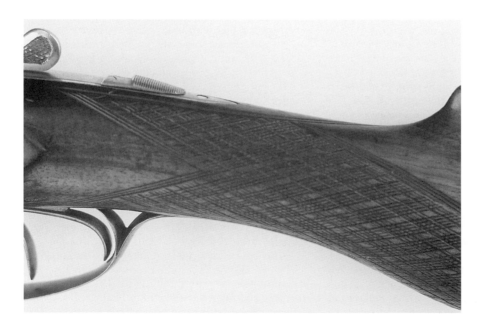

Skip line chequering, rather unusual on a side-by-side.

Gunsmith-made chequering tools. Note how a tool with a wider line spacing has coarser teeth. This is sometimes done quite unconsciously when making the tool.

Chequering has evolved from square-patterned wide-cut lines (sometimes with a dot in the middle of each square) to finer diamond flat-top patterns, and finally to the diamond pointed generally in use today. The spacing between lines gradually became narrower as finer patterns became the norm. A late percussion gun, for instance, may sport some fourteen to sixteen lines per inch (lpi), which gives a very good grip and is well suited to the heavy hand of a muzzle loader. On a later breech loader of slimmer stock dimensions, it is more usual to find chequering of twenty-four lpi. Occasionally guns were made with chequering as fine as thirty-two lpi, as a kind of *tour de force* of the chequerer's skill, but when it gets this fine it is in danger of no longer serving its original purpose.

Some makers, particularly Greener, stayed with the flat-top diamond pattern after other makers had changed to diamond pointed. Really good flat-top chequering may look a little plainer than diamond pointed but is actually more difficult to achieve and a greater testimony to the chequerer's skills. Also, chequering should be in proportion to the size of stock, and an eight-bore wildfowling gun will normally carry heavier chequering than a dainty twenty-bore game gun. For example, Webley and Scott would use twenty-four to twenty-six lpi (and sometimes finer) on their game gun doubles, while on the Greener GP they continued to produce eighteen to twenty lpi, visually much better suited to the thicker hand of a Martini action stock.

Skip-line chequering, where some of the lines are not cut across to produce a diamond, gives a pattern within a pattern, and fleur de lys borders are most often found on mass-produced guns of American origin. Such designs are a matter of taste, but really ill suited to the slender stock and forend of a side by side double gun.

Methods of Chequering

Traditionally, chequering is cut by hand with a small tool having two parallel lines of teeth. There are also single-cut tools, useful for recutting or border work. With such simple equipment both flat-top and pointed diamond chequering can be produced, and a good chequerer will do the job with most of the finish already on the stock. This gives a cleaner, crisper appearance, as finishing after the chequering is cut tends to lead to oil clogging at least some of the border, which will require recutting.

Chequering is one of the features of a gun that everyone notices. It is not a job for the fainthearted and requires a peculiar skill that eludes many gunsmiths. Cutting lines around quite complex curves and making them appear straight on a stock that is ninety-five per cent finished is quite a job. Fortunately, repairs to chequering are rarely extensive and much of the pattern already on the stock or forend can be followed.

Other methods of chequering are produced by machine. These include pressed-on chequering, machine-cut and, more recently, laser-cut chequering. While the gunsmith is not going to have the equipment to replicate these methods of production, most repairs, with some skill and a little cunning, can be effected by hand.

Cleaning Chequering

When I was an enthusiastic youngster, a contemporary advised me to clean out chequering by brushing it very gently with a worn toothbrush. It did not take me too long to find out that he knew no more about the subject than I did, as this method does not work. Chequering at the bottom of the cut is narrow, even pointed, and the splayed and rounded bristles of a worn toothbrush are not suited to cleaning into these places.

He was right in one respect, though. Chequering that looks worn and dull sometimes needs only a good clean to restore a bolder, tidier appearance. With handling and exposure, the cut lines become gradually clogged by dust mixed with grease from the hands. Sometimes this dust is quite gritty and, when recutting with a chequering tool, it can lose its sharp edge quite quickly.

Before entertaining any idea of recutting, it is best to clean out the chequering thoroughly; this may be all that is required to enhance its appearance. Any toothbrush that is used for this job needs to be in good condition – fine and stiff-bristled – and used with either warm, soapy water or a spirit that will soften the hardened deposits trapped in the chequering. Brushing in white spirit will work on these deposits, and finishing with warm soapy water will remove the spirit smell. Another benefit of using soapy water is that the fine bubbles formed from the brushing will change colour as they become contaminated with the dirt and grease. When the bubbles no longer change colour it indicates that the chequering is properly cleaned out.

These methods work particularly well with pointed diamond chequering, but flat-top chequering is more of a nuisance, and a better tool for the job is a very thin-bristled brass suede brush. These are harder on the wood than nylon and care should be taken to stroke the brush along the chequering cuts rather than using the scrubbing motion that is acceptable with a toothbrush. Often it is necessary after drying to do some minor refinishing, and a thin application of Tru-oil or French polish direct from the ends of the fingers, to cover the diamonds without running into the chequering cuts and defeating the object of cleaning, is all that is required.

Recutting Chequering

Recutting involves going over existing chequering and cutting it to a uniform depth. For recutting, a single-cut tool can be used very effectively on flat-top chequering, but the pointed diamond form, if worn, requires a tool with two sets of parallel teeth, to give the correct diamond shape.

It is easiest to start where the existing chequering is unworn and work into those parts of the chequering that are worn, using short cutting strokes in a backward and forward motion, almost like filing. This is where the double-cut tool is an advantage as it has less inclination to slip across into another cut and spoil the line. Some double-cut tools are made with one cutting edge deeper than the other. The deep side follows the existing line, while the shallower edge marks out the new line. After that, it is just a matter of moving the tool across to repeat the operation. Once the panels are recut right up to the edge, the border can be recut and it looks better done just a little deeper than the rest of the chequering. A narrow-angle single-cut tool is especially good for this job. Borderless chequering is a difference proposition. It is not, of course, actually borderless, as this would not be practical. What it means is that there is no contrasting border and all the cuts should be to an even depth.

Cutting in Repairs

This includes wood let in to the side of the chequering, not uncommon with a damaged splinter forend, or a patch let into a chequering panel, either as a structural repair or just to fill in an area of damaged chequering. When new wood is let in it forms a step against the edge of the existing chequering and it can be tricky to form the start of a clean cut. It is best to make a 'lead in' in the form of an angled cut from the existing chequering line into the new wood. This can be done by cutting from the new wood towards the existing cuts with a chequering tool, but is most easily accomplished with a needle file made for chequering use. Once the 'lead in' has been produced, the cut with the chequering tool can be made from the existing lines into the new wood.

An alternative to the needle file is a veiner tool, a short-shafted, vee-shaped chisel with a large handle to fit the palm of the hand. It takes a little more care and skill to use than a needle file but also has its uses when cleaning up cuts at border junctions.

Stock Bending

Introduction

Stock bending is akin to spill boring in that it involves a certain level of mental tension, fostered by the knowledge that encountering an unexpected, unseen fault could be the ruination of the stock. Certain precautions can be taken to minimize the risk but, without the benefit of comic-book x-ray vision, it is not possible to see into a stock to examine the structure. When we used to buy three-year naturally seasoned walnut in large slabs and cut our own stock blanks, it was amazing how much tension would be in the wood, sometimes releasing with quite a bang as the blanks were cut out. The blank would gradually change shape, twist and deform, and, even after a further minimum of two years in store, could hold further surprises. The worst problem was shakes – small cracks in the grain that went nowhere, but usually occurred in an awkward place. The timber supplier would always respond to reports of problems with a shake of the head, and the comment, 'That's wood for you. Always a bit of an unknown.'

He was right. Wood is always unpredictable. There is far less risk of problems on a good-quality gun in decent condition, with specially chosen fine walnut, than on a cheaper gun, where the price of the wood is a greater consideration than its quality or, maybe, even its fitness for purpose.

Cast Off – Cast On

Gun stocks are not straight (or, at least, should not be straight) – mounting a gun at the shoulder and looking in the same direction as the barrels is

The terms 'cast off' and 'cast on' (in other words, away or towards) must have been devised by a right-handed person. It is the only logical explanation.

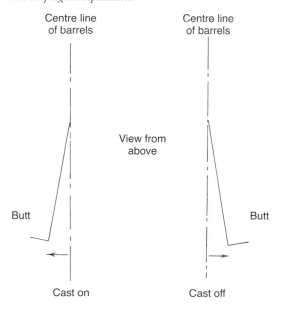

almost impossible with a straight stock. It can be achieved in a static environment by pushing the head over at an unnatural angle, but for moving targets this is not feasible, and the gun has to fit the physique of the user.

Stock bending is primarily concerned with cast and drop, and cast is either 'off' or 'on'. Looking from the butt forward to the barrels, cast off is with the butt to the right of the gun's centreline, and cast on to the left of same. Complementary to the cast is the position of the toe, which should be even further over than the direction of the butt as a whole and, in extreme cases, can give a slightly strange, twisted look to the stock.

Drop is the difference in measurement between the tip of the comb and the heel of the butt relative to a line across the barrels. All these combinations of measurement can have a dramatic effect on a shooter's performance and, therefore, on his or her enjoyment of a day's sport. A vast number of shotgun shooters shoot with ill-fitting stocks, and the only reason so many shoot reasonably well is because of the remarkable, and sometimes almost unconscious, human ability to adapt.

With experience it is possible to 'guestimate' quite accurately the stock cast for a particular individual's physique, but by far the best method is to suggest a session with a try gun at the pattern plate and a few clays. A local shooting school may provide these facilities, and customers will be given a card detailing not only cast and drop, but also length of pull and any other details the coach may feel relevant. All this information will be very useful to the gunsmith.

Taking Precautions

Prior to setting up for bending, the stock should be examined both externally and internally. Short grain in the hand – in other words, the grain running diagonally across the hand of the stock rather than in line with the stock – is a primary reason for rejection. It is possible to bend such a stock, although the risk of breakage is high and, particularly with an over-and-under, the chances are that one side of the stock will crack along the grain after comparatively little use.

Any external cracking, which most often occurs around the end of the tang or top strap, is certainly cause for turning down doing the job, as are any signs of stock repair.

With the stock detached from the action, check visually for signs of cracking or repair. Sometimes a crack along the grain is not a problem, if it has been caused by the stock being bent previously and the

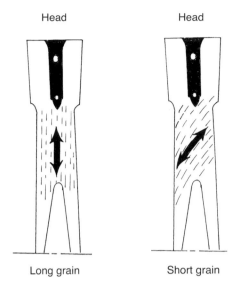

Head   Head

Long grain   Short grain

There is little sense in trying to bend a stock with short grain through the hand. There is even less sense in using such a piece of wood for a stock, but it is done.

Methods of Bending

The two common methods of bending are steaming or hot oil, and the latter is probably the most widely used by the gunsmith. Steaming is much favoured for bending wood in general woodworking, but the disadvantage on a finished stock is that it may raise the grain and spoil the finish. With either method, the principles are the same. The stock, when heated by the application of hot linseed oil or steam, becomes soft, and can be moved without breaking and left to 'set' in another position.

It is normal to use a jig to accomplish the bending with ease and accuracy, although until recently one famous Birmingham maker used, with remarkable effect, a broom stale between the window ledge and the stock and action held in a vice. Jigs are either screw- or wedge-operated and the screw type is available commercially. I prefer a wooden jig, using wooden wedges driven into place with a wooden mallet. It is not so quick or convenient to set up as the proprietary screw jig, and it can be trickier to get the stock set accurately to the new dimensions. However, it has one great advantage over the screw type: it is possible to detect when the wood is ready to move, by hitting the mallet against a wedge. Before the stock is ready to bend, the result is a sharp wood-on-wood tapping sound. As the stock becomes softened and pliant, the sound will change to a dull 'doff'. It is then safe to apply

intention is to bend the stock back to the position it held prior to the first bending. In such an instance this 'stress crack' will often close up as the wood goes back into its natural, stress-free position.

My stock-bending jig – saved from a wood pile and much modified over the years. To go with it is an ammunition box of various clamps and wedges, all made from scrap wood.

pressure without stressing or possibly cracking the stock – a comforting thought!

Setting Up and Bending

The jig I use accepts the whole gun, which lies upside down along a centreline cut into the base of the jig. Clamps are arranged at intervals to hold the barrels true to that centreline and also clamp the action in place. Surrounding the butt is a bridge piece into which wedges are placed to alter the cast and/or drop of the stock. Beneath the hand of the stock the base of the jig is cut away, and under this is a container of linseed oil with provision for heating.

Where the hand of the stock is to be bent it is useful to wrap it with string (the brown, loosely woven garden type is particularly good), as this concentrates the heat of the linseed in that area. A couple of loose ends hanging down also act as a run-off back to the linseed pot.

NB: *Great care should be taken when heating linseed oil. It should only be simmering and never brought near the boil. It is quite unnecessary to have it boiling hot, and linseed will, when it reaches a certain temperature, boil uncontrollably, like milk. Burns from contact with hot oil are particularly unpleasant, so you should also protect yourself from splashes and spillage by gloves with long sleeves, stout footwear, eye protection and, if available, a leather apron such as those used by welders and blacksmiths.*

With the oil simmering nicely below the stock, a soup ladle with a pouring beak is ideal for applying it to the wood. Usually a minimum of thirty-five to forty pourings are sufficient to prepare a slim stock for bending, but individual pieces of wood vary, and it is always better to apply too much rather than too little. Heavier stocks with pistol grips, and especially over-and-unders, require many more applications with the oil before they are ready to move. Once that dull 'doff' sound is heard on the wedges, they can be carefully driven into place to give the required dimensions.

When resetting a stock it is advisable to slightly exceed the measurements you wish to finish up with, as the wood will often move back a little way within twenty-four hours of the bending. Once again, the unpredictability of the wood comes into play, and being able to make an educated guess at how different pieces will behave will come only with experience. As a simple guide, if a stock bends quickly and easily, there is a good chance it will stay pretty much where it is set. If it takes a lot of heating and still shows some resistance to being reset, there is an equally good chance it will move back

an indeterminate amount.

As soon as bending is complete, the heating for the oil should be turned off, the string removed from the hand of the stock – it will be hot – and all surplus linseed removed with a dry cloth or paper towel. The gun should be left in the jig for several hours until fully cooled. Ideally, stock bending should be the last job of the day and left in the jig until the start of the next working day. The stock will then need to be stripped off the action again to clean out any surplus linseed oil. If this is not done, the linseed oil will almost certainly 'glue' a tang safety in place and, with a side lock, can even slow the fall of a hammer and make the sear decidedly sticky in operation. After a clean and lubricate – a mini-service, in fact – all that remains is to fit it back together.

Minor Problems with Stock Bending

Sometimes the hot oil will affect the stock finish or appearance. On a side-by-side the bending takes place where there is a lot of chequering, and linseed will enter the stock, sometimes producing a darker area in the wood. This will usually fade and disappear after a couple of weeks.

Varnish stock finishes may try to lift away from the wood and this potential problem should always be pointed out to a customer. When it does happen, a repair is the only option.

Guns fitted with a stock bolt, such as most over-and-unders, are limited in the amount of bend that can be produced due to the nature of their construction. They can, however, be bent and sometimes opening out the hole through which the stock bolt runs can be beneficial. It is important to be careful of one thing with these types of gun: after bending, check the fit where the sides of the stock head up behind the action. These should be even, but the effect of bending can produce more load on one side of the stock. If this happens the side making hard contact should be relieved until it fits properly, otherwise adding the jarring impact of recoil to the load already applied by the stock bolt can result in breakage on the one side of the stock.

Problems not Cured by Bending

While a stock with the correct amount of cast, drop and length for a particular individual is a quantum step in the right direction, it is not always a 100 per cent cure-all. The combs on many over-and-unders are very rounded and wide so, even with the correct cast, sometimes either the gun cannot be mounted properly without the user pushing his head hard against the stock, or it smacks against the cheekbone under recoil. The answer is to offset the comb from the line of the butt in the same direction as the cast. This can be done by subtly hollowing the stock and producing a saddle effect rather than moving the comb in a straight line.

This problem is not confined to modern over-and-unders. It is not unusual to find older side-by-sides that were made and fitted to a particular customer displaying this feature, albeit done so cleverly it is not immediately obvious.

I once received a pair of guns made in 1898 and noticed that, not only did they have long stocks and a lot of cast, but the combs were curved, and a line taken across them ran almost parallel with the centreline of the barrels. Years before, I had known some of the older members of the family that owned them, and their common physical characteristics were broad shoulders, long arms and wide, square-jawed faces. Of course, this pair of guns fitted them quite well.

A William Powell crossover stock – a superb example of the gunmaker's work. While it has a slightly odd appearance, it is an amazing creation, with the back of the action and even the lock plates curved to follow the lines of the stock.

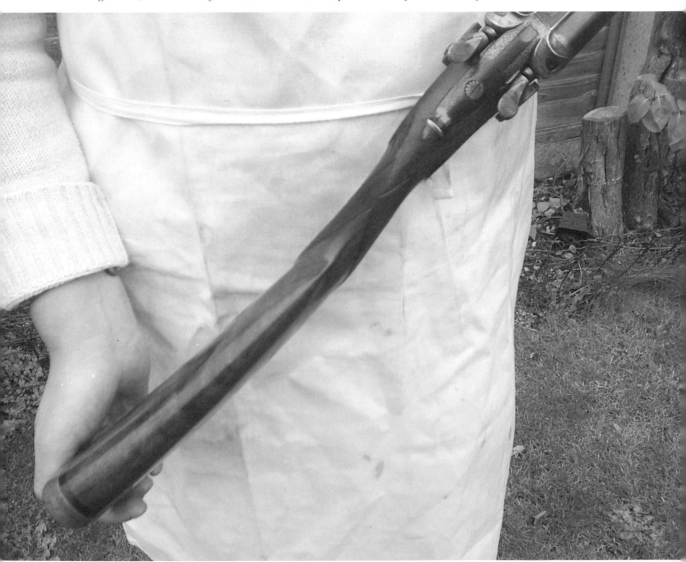

It seems hard to believe, but the length of comb can also have an effect on the middle finger. If the comb is too far forward it will hold the hand in such a position that this finger is pushed into contact with the trigger guard, sometimes resulting in bruising. Occasionally trigger guards have rubber buffers wrapped around the rear of the bow, but this is not the answer. The cure to the problem is to move the comb back so that the hand sits in the correct position.

Cross-Over or Central-Vision Stocks

These seemingly extravagant stocks are made so that the user can mount the gun using the opposite eye to shoot, or to get a better balance between each eye. Normally these would be used by, say, a right-handed shooter who had impaired vision in the right eye, for whom shooting left-handed was not a satisfactory option. While they look rather odd, good results are quite often obtained, even if the user does not have really impaired sight. For example, where a shooter has a strong opposite master eye, a cross-over stock can seem an advantage, but whether they actually need a cross-over stock is a different matter. Whether they are necessary or not, they are none the less rare and interesting examples of the gunsmiths' art.

It might be possible to produce an approximation of a cross-over stock, but it would require some drastic bending, which could only be achieved by considerable steaming. Even then, the results would not be completely satisfactory. On a true cross-over gun, the back of the action is biased to one side; the top strap, trigger guard and the lock plates, in the case of a side lock, are all curved. Cross-over stocks can be made for over-and-unders. As this would interfere with the run of the stock bolt, a form of 'thumbhole' stock with a cut-out section in the pistol-grip area is one practical solution, if visually rather unusual.

CHAPTER 11

Blacking, Bluing and Browning

Introduction

Traditionally, fancy and protective gun finishes were brown rather than black, and originally it was a simple form of controlled rusting, which helped prevent later corrosion. It is alleged that the 'Brown Bess' musket of British military fame, which saw stalwart service at Waterloo, was thus named because, as issued, it had a browned finish to its metalwork. Browning is still used for Damascus barrels as a controlled rusting process, not only for protection but also to enhance their beauty. Blacking is used for steel barrels and the furniture and bluing – meaning heat bluing – for pins and small parts. Early blacking was oil blacking produced by heating the component and dropping into oil, which, if it was a little dirty (containing carbon), only enhanced the durability of the finish.

It is not uncommon for a customer to ask for his shotgun barrels to be 're-blued', but this is a term normally reserved for the processes carried out on rifled firearms, such as heated charcoal bluing. The far more prosaic-sounding blacking process can be cold blacking, another controlled rusting process, or express blacking, a hot-water method. There is also caustic blacking, which is unsuitable for shotgun barrels, but superb for shotgun furniture and rifled firearms.

Handles are essential for removing barrels from hot water. The arms can easily be made from welding rod and the handles are better if they do not revolve, hence the idea of bonding them in.

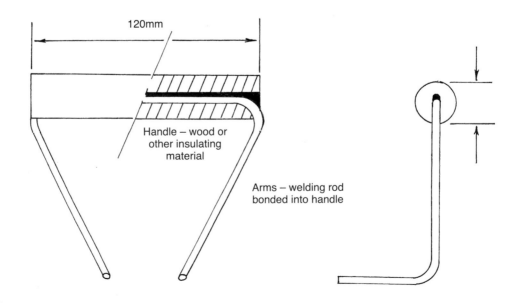

120mm

Handle – wood or other insulating material

Arms – welding rod bonded into handle

Heat bluing as normally practised does not give any corrosion protection but is used to provide contrast with mating parts. The pin heads inside a lock are a favourite for this treatment, then providing a classically colourful contrast to a polished bridle.

The definitive publication on these processes is R. H. Angier's *Firearms Bluing and Browning*, first published in 1936 and reprinted many times. It is such a standby to the trade or hobbyist, that it is known simply as 'Angier's'.

Express or Hot-Water Blacking

The Method

The most common method of blacking shotgun barrels is the express or hot-water method. This requires two things: carefully degreased barrels and firm ribs that do not leak air. Many imported guns have an air hole in a rib, and this has to be blocked, perhaps by the shaped end of a matchstick.

Barrels that have been polished by mechanical means using a polishing compound are almost impossible to degrease adequately for this method of blacking. They should be finished by hand using wet and dry paper of 800- or 1200-grade, depending on the desired final finish. There are some advantages in using the slightly coarser-grade paper, as the blacking solution seems to adhere better if the finish is not too smooth.

Degreasing is absolutely essential and can be done as a two-stage process: first, scrub the outside of the barrels with hot water and plenty of washing-up liquid, using rubber gloves to avoid transferring greasy finger marks to the cleaned metal; then dry with a soft cloth.

The full degrease is done by one of the most quaint but superbly efficient methods ever devised – and still far better than most methods of chemical degreasing. A long, narrow tank is filled with enough soft rainwater to cover the barrels and set up over a gas burner (or, as an alternative for a commercial operation, with an electric immersion heater. This requires a deeper tank and protection above the heating element so that the barrels do not make contact.) Rainwater is used because it is nearly pure or, at least, chemical-free. The water only needs to be simmering when the barrels are lowered in by means of handles and left for a few minutes to heat up to match the water temperature.

While this is ongoing, a mixture of chalk dust and water is mixed in a shallow bowl (a saucer is big enough) to form a slurry about the consistency of emulsion paint. The barrels are removed using the handles, laid on a bench covered with newspaper and, as soon as the water has evaporated, painted with the chalk slurry using a small paintbrush.

Within seconds the chalk slurry will start to dry, usually from the breech ends first, as they retain the most heat. When completely dry, the chalk is removed with a degreased suede brush or '0' grade wire wool. It is convenient at this point to have more permanent handles to hold the barrels; long tapered wooden plugs that can be gently pushed into either end suffice for this purpose. The barrels are then immersed in the hot water again, and the process is repeated until, on being lifted out of the water, the barrels remain wet until the water evaporates. This indicates that they are fully degreased. If the procedure is done properly, and with a pre-degrease as described, often the barrels only need to be 'whitened' three times to obtain the final result.

There are two schools of thought as to whether at this point the tank should be emptied and filled with clean water for the blacking, or whether it is acceptable to use the water that is slightly contaminated with chalk and perhaps traces of grease. In the absence of a second tank already set up, the former will avoid any potential problems later on, although some trade professionals seem to be able to use water the colour of a river in flood with superb results.

Whitened barrels, one of the best forms of degreasing prior to blacking. In this case the barrels were being browned and accidentally got caught with a spray of oil. Several applications of chalk and hot water removed the oil, allowing the browning process to be continued. Note the tapered wooden plugs, which make handling easier.

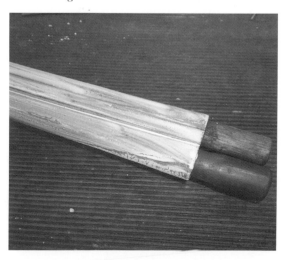

The next move is to prepare the blacking solution. This can either be used cold or hung in a container at the end of the tank to be warmed. The disadvantages with the latter is the rate of evaporation of the solution, which is expensive and commercially not very easy to obtain. A small piece of worn cotton sheet tied to a thin hardwood handle – a simple form of brush – makes a good tool for applying the blacking solution to the barrels.

NB: *These chemicals are poisonous and the manufacturer's instructions should be strictly adhered to. This is not a process for doing on the kitchen stove. Protective rubber gloves are a necessity and the blacking solution should be kept in a locked cupboard when not in use.*

When the barrels have been heated in the water and then laid on the bench, with the wooden plugs in place, the blacking solution is applied after the last of the hot water has evaporated. Using the minimum amount on the applicator, it is brushed on using long, even strokes for the full length of the barrels and left to dry, which can take less than a minute. The barrels are then re-immersed in the water to heat up.

When the barrels are removed, the film of dried-on blacking solution will look quite dark, but this now has to be brushed off – or 'carded off', in the old terminology – and only a trace of colour change will be left. This can seem a little disappointing at first but after several applications of the blacking solution the finish will start to build up. About eight repeated applications will give a good finish and ten to twelve will do that luxury job. After that, the results will not improve any more.

After the final treatment the barrels should be placed on end and allowed to drain. Blacking will have got into the bore(s) but this is not a problem as it can be polished out later when the lumps are cleaned up. However, it is important to put an oiled patch through the bore(s), preferably while still warm, to avoid corrosion, and to wipe over on the outside with an oil-dampened rag after cooling. Before refitting to the gun, both muzzle and breech ends of the barrels will need to be polished and the extractor(s) done to match for a professional job.

Problems with Express Blacking

Streaking is caused by inadequate degreasing and this becomes obvious on the first application of the blacking solution. The only answer is to start the degreasing process again.

Spewing is caused by loose ribs where air escapes from between the ribs, water enters and the result is a patch where the blacking will not take. There is no answer other than to re-lay the rib(s).

A soft finish to the blacking, where it looks good but rubs off easily, may be caused by not having the water hot enough or allowing the barrels to cool too much before applying the blacking solution.

Sometimes, mainly with late manufactured barrels, shiny areas will seem impervious to the blacking. Applying hydrochloric acid in fairly dilute form to the warm barrels will, if left about an hour, change the colour of the metal's surface. Starting the blacking process again after this treatment will often get rid of the shiny areas.

Sometimes it is suggested that plugging the barrels for immersion is a good idea but is really just a complication that is not necessary and rarely used by professionals.

Caustic Blacking

The point should be made right from the start that caustic blacking is not for shotgun barrels. If barrels with soldered-on ribs are put into this solution, the solder will be quickly eaten away, and the ribs, forend loop, and so on, will all be nicely detached and solder-free when the barrels are lifted out of the tank. Sometimes barrels with silver-soldered or brazed ribs are blacked in this manner but, although the ribs will stay in place, it is a savage process and will penetrate even the tiniest pinhole between the ribs. Later the salts will start to grow out, sometimes for many months afterwards, and they are very caustic.

However, for shotgun furniture such as top levers and trigger guards, caustic blacking is a most convenient process that gives an excellent finish and good protection against rusting.

The Method

The salts come with most comprehensive instructions on safe handling and use. It is then a choice of either setting up a tank at what seems a convenient working height, as with the express blacking, or on the floor. If the tank is on the floor, and against a wall, it cannot be tripped over; a tipped-over tank and a lap full of caustic solution heated to over 140 degrees centigrade does not bear thinking about.

When adding the salts to the water it is essential to use a full face mask for protection as well as long rubber gloves, and these should also be used when the process is in operation. Care should be taken when initially heating the water by a gas burner that the salts do not settle on the bottom of the tank, so it is necessary to stir until dissolved. Once

A caustic blacking set-up, with long and short tanks, a perforated container for small parts and a bucket for hot water for swilling off. The gas bottle is not ideally situated. Later it was moved outside the building and piped in using rigid copper tubing.

the solution is ready, it will boil at 143 degrees centigrade in what is accurately described as a rolling boil. A thermocouple is a luxury, not a necessity, and is only any good if it is correctly calibrated. With a little practice it is possible to judge whether the mixture is correct and this can be backed up by using a test piece hung in the salts before blacking the actual job parts.

The parts should be degreased but as the salts are very caustic a wipe over with a rag damped with methylated spirits is sufficient. This reduces the amount of contamination in the tank and prolongs the life of the salts. The surface finish of the parts to be blacked is most important and relates directly to the finish. A highly polished, almost mirror finish will produce the deepest black. A polished part acid-etched will come out as a satin black and parts very finely bead-blasted will finish up matt black.

Once the test piece has blacked satisfactorily, the gun parts can be lowered in on thin wires – plain iron or stainless, not plated – and fully submerged, otherwise a dirty 'tideline' finish will be the result. When, on examination, the parts are fully blacked, it is worth agitating them in the tank and leaving for another minute or so for the final finish.

After withdrawing from the tank the parts should be flushed in clean water – swilled on the end of the wire in a bucket of clean water. This should be followed by oiling with a water-soluble oil, then leaving to drain, preferably until the next day.

Problems with Caustic Blacking
If the solution is too strong it will burn on the steel, spoiling the finish, and the parts will emerge a dull reddish-brown.

NB: *Adding water to a strong caustic boiling solution is a dangerous move; the burner should be turned off and the solution allowed to cool before diluting. It is best to start with a slightly weaker solution so that, as the water evaporates and it reaches its working temperature, there is some time to do the work before it becomes too strong.*

As the solution is reused it weakens and becomes contaminated. Eventually it will take a long time to produce a blacked component, which may emerge with a brownish deposit. In this instance it can usually be rubbed off after flushing with clean water using soluble oil and '0000' grade wire wool. By the time the salts are this poor it is time to make up a new solution.

Some of the most unpredictable parts are shotgun trigger guards, which may emerge from the solution with spots of startling brightness where the steel has remained untouched by the blacking salts. Most of the time a wipe over with acid will cure this problem. If not, the part can be lightly pickled in acid, but this means it will come out more of a satin black, which, in itself, has a not unpleasant appearance.

Browning Barrels

The Method

Browning – the controlled rusting of ancient lineage that brings out the beauty in Damascus barrels – is rather a gentlemanly pursuit. Unhurried but very satisfying, it can be done quietly in between other jobs.

Browning solution can be made up from one of Angier's 'recipes' or can still be obtained commercially in small quantities. As this is a simple process, it is best to fit wooden plugs that will stay in place for most of the job. After degreasing the polished barrels with methylated spirits, the first application of browning solution is best made using a small pad of fine wire wool.

NB: *Many browning solutions are poisonous, even when used in small quantities. Rubber gloves must be used. If it is done first thing in the working day, the barrels can then be left propped up to start rusting. If, a little later, the finish appears streaky, denoting poor degreasing, it can sometimes be corrected by applying a little more browning solution with the wire wool and blending in with the areas already rusting. If this does not work it means repolishing and degreasing again.*

Once an even layer of rust is formed, the pattern in the barrels will start to become visible. At the end of the first day the rust can be rubbed off with '00' grade wire wool and the pattern will still be just visible on the barrels. Changing now to a small piece of cloth to apply the browning solution has the advantage of using less solution and being able to apply it without dribbles. After the barrels have been left overnight, the rusting can be rubbed off again the next morning and more solution applied. After the first four or five applications the rusting process will usually slow down until the solutions needs applying only once a day.

When it is judged that the pattern is well defined, the iron in the barrels is quite dark and the steel brown, the barrels should be boiled in a tank of water. Logwood added to the water will impart a dark plum colour to the brownness and help neu-

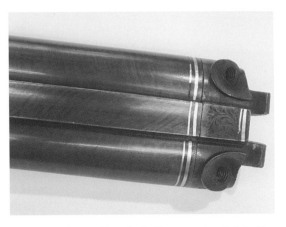

The finished browned barrels of a Damascus-barrelled double – a pleasure to the eye!

tralize the finish to prevent after-rusting. A few washing-soda crystals added to the logwood solution will aid this.

After removal from the logwood solution the last layer of fine rust can be gently removed with '0000' grade wire wool. When dry but still warm, the barrels are wiped over with some linseed oil on a cloth, which will impregnate the browning and dry like a varnish to enhance the finish. The barrels can then be put aside somewhere warm and dry for a few days before polishing the barrel breech and muzzle ends, not forgetting to oil the bore(s).

Problems with Browning

There are few problems with browning; the most common is after-rusting, which occurs because the acid-based browning solution has not been neutralized. Leaving a solution on too long before rubbing off can result in a dense, hard layer of rust that has to be cleaned off with a fine file, meaning the barrel has to be struck up and polished/degreased to start the process all over again. The area where the ribs join the barrels is difficult to get into and, to avoid a build-up of rust, a suede brush can be used, also around the foresight bead.

Heat Bluing

I was once told that, to obtain a stunning heat blue, all I needed to do was rub over the components with whiting (chalk), put them in a sealed tin of charcoal and leave it in the workshop coke stove overnight. I tried this several times and managed to produce some amazingly varied shades of grey/blue, but nothing worth showing off. I must have been missing some part of the secret, so I went back to using the propane torch.

Using the torch is simply applying the knowledge that, as steel is heated, it changes colour and that, of course, is long before it starts to glow! The sequence of colours is as follows:

- pale yellow;
- straw yellow;
- golden yellow;
- brown;
- brown with purple spots;
- purple;
- bright blue;
- full blue;
- dark blue;
- followed by just starting to glow red.

Parts such as pins for locks are too small on their own to be able to control the temperature and, therefore, the colours. A small steel block with holes to accept the pin bodies, placed on a firebrick, will give a reasonable heat sink and more control. Heating the area around the block with a propane torch rather than directing heat on to the screws is best. The colours will appear relatively slowly and seem to wash over the slotted pinheads, so they can be stopped with a spray of gun oil while holding the propane torch well out of the way.

The only problem with heat bluing is going past the colour you want, and then repolishing after cooling and starting again is the only option. The other small point is, although the method is referred to as 'bluing', purple makes an attractive colouration.

CHAPTER 12

Engraving

Introduction

Engraving is a specialist skill carried out by someone who not only has mastered the required techniques, but is also an artist. It is one of the jobs that the gunsmith invariably sub-contracts – certainly, if it involves a few letters or, perhaps, overall coverage – although it is not beyond the gunsmith to make simple patterns on screw heads. Most tang screws, for example, are engraved in a simple style that deceives the eye, making the brain assume more than is actually there.

A small hand graver (engraving tool) and, especially if your eyes are not in the first flush of youth, magnifying glasses with binocular vision are necessary to get the best results. Using a knife-edge file is not an option. There is a distinctive difference between a file cut and the clean-edged line produced with an engraving tool.

It is something worth trying, and if a job is urgent it can save time to decorate a screw head yourself, rather than sending it to an engraver, with all the potential delays of postage, and so on.

For some basic engraving, just one simple tool will suffice – and, for me at least, extra optical help for the close-up work.

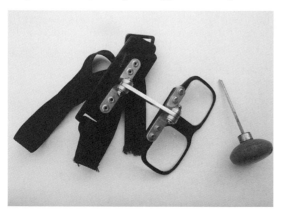

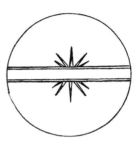

First cuts deeper towards slot

Second cuts deeper towards outside

Third cuts infill – same depth

Simple decoration on screw heads may be designed to deceive the eye but is none the less very effective. Taking a cut from the outside towards the screw slot enables the operator to deepen, and therefore broaden, the cut in one stroke. The same applies working from the middle outwards.

Gilding the Lily

Engraving varies enormously, from examples of subtle and restrained beauty, via those of simplicity and modest coverage, but fine quality, to some that might be described as 'in your face', that is, rather obvious and over the top, or stunningly beautiful, depending on taste.

Many of the more elaborate examples, usually found on collector's guns, represent the best engraver's skills, with the sought-after features of relief carving and lavish gold inlay. Much of the engraving is a work of art in its own right; indeed, sometimes the gun seems to be merely the carrier for the decoration. Such exoticism, something of a parody of the more extravagant continental styles, seems at odds with the restrained style of the typical British shotgun, with its elegant lines and handling, and its tasteful decoration. Of course, it may not be what the gunmaker would prefer to produce, but he has to be led by the customer.

There are still plenty of examples of fine engraving to be seen, carried out in a variety of styles and techniques. There are lavishly illustrated books on engraving, and other useful references include the glossy magazines, and the better auction catalogues, where a variety of guns, from muzzle loaders to modern-day firearms, are illustrated, and types and styles of engraving can be studied.

Within these pages terms such as rose and scroll, beaded border, best bouquet and carved fences come to life and start to have real meaning.

Methods of Engraving

Of those patterns seen on lock plates and actions, not all are actually engraved. In fact, of the current processes used, only one method qualifies as engraving: the removal of metal with a sharp hand tool and cutting out a pre-determined pattern on the surface of the steel.

For many years this was the common method of decorating guns. There were plenty of trade engravers competing for work and even some of the bottom-of-the-range late-production hammer guns display some hand engraving, although this is sometimes little more than a border line and a name. Game scenes would vary from stunning artistry at the top of the range to economy jobs – sometimes referred to in the gun trade as the 'Donald Duck School of Engraving' (that is the polite version), because of the Disneyesque appearance of the wildlife. This type of engraving often consists of just a few cunning lines designed to deceive the eye and leave an impression of a pheasant or duck. It should not be confused with some quality engraving where the animals appear somewhat odd, but where the engraving is fully detailed and finely

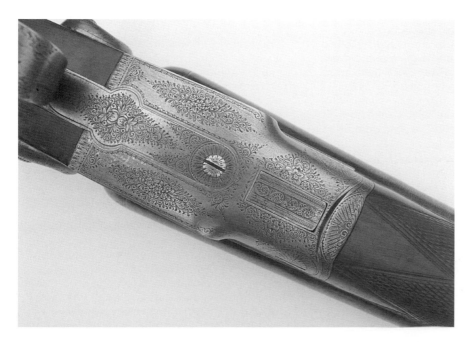

Fine rose and scroll engraving on a late nineteenth-century double. This was a very late hammer gun, whose original owner did not trust the new fashion for hammerless guns.

Work from the 'Donald Duck' school of engraving.

executed – the late nineteenth-century lion, for example, sometimes looks like an amiable teddy bear sporting a bad wig. It looks strange to modern eyes partly because of changes in artistic fashions, but also because the engraver would quite possibly never have seen a real lion.

Hand engraving requires the development of a design, which is then cut out using tiny chisels of different shapes. If a background to a scene is to be removed, it is dug out one sliver at a time. Deep cuts are made using a small hammer to drive the chisel. Small cuts are made with a hand tool, which might have a tip no wider than the point of a needle. All cuts are characteristically sharp, bright and light-reflecting. To emphasize the subject each cut may also vary in width and depth. Depending on how the work is viewed in relation to the light, it produces a variation in light and shade on the engraving, giving a powerful contrast to the work. If the work is tilted, the cuts will shine like waves across the surface and this gives the hallmark richness that is only produced by true quality.

Nowadays, most hand engraving is reserved for individual orders or bespoke guns. More economic methods of producing decoration on mass-produced guns include etching, stamping and machine engraving.

Etching is basically a process where acid is used to selectively eat away parts of the metal to produce a pattern. Protection of those parts to be left bright is achieved by coating with an acid-resistant substance. The resistant coating can be ready-printed if the pattern is to be the same on a number of pieces, thereby increasing efficiency and reducing costs.

Industrial electro-etching works in a similar manner. An electrically charged pad carrying an electrolyte operates through a special stencil. This method is sometimes used to print non-decorative information, such as lettering and serial numbers. While quick and very precise, the indentation produced is rather shallow and will not stand up to much refinishing.

There is also decorative etching, where acid-resistant waxes are applied to the metal and the desired pattern is cut by hand through this protective medium. The work piece is then submerged in an acid bath and, where the metal is exposed, it is eaten away by the acid. After halting the process by rinsing and then dissolving away the wax, the original polished surface is uncovered, with the etched pattern now apparent. This process may be repeated to produce sufficient contrast and detail, and highlighted with hand engraving. If gold inlay is added, it is possible to produce some visually stunning work. Such a combination of techniques still requires the hand of a true craftsman and artist.

Impressed or stamped patterns may be pressed or rolled into the metal surface by machine. In the nineteenth century the cylinders of some Colt percussion revolvers were rolled with a suitably inspir-

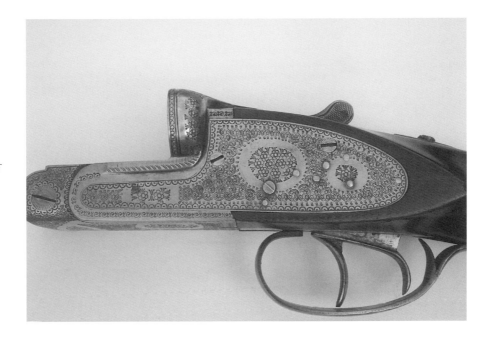

Stamped-on imitation engraving – simple, even possibly crude in terms of engraving but, to many, quite effective.

ing scene; probably the best known is the Colt Navy revolver. Today this technique is still used when producing reproduction Colt revolvers of that period. It has also been refined to an extraordinary degree to produce the appearance of engraving on mass-produced over-and-under shotguns.

Machine engraving is seldom used in gun work due to its relatively coarse nature and inability to produce fine detail. However, it can be used for adding initials to brass gun-case labels.

Identifying the processes used to decorate a gun is sometimes complicated, because different methods may be mixed in one job. Impressed decoration leaves a slight raising of metal on each side of the impressed line. Because no metal is actually removed, it is displaced, and a slight bulging is created. This can sometimes leave a visual impression of the pattern being slightly proud of the surface. However, this method can be tidied up and then enhanced with hand engraving to increase the detail. Hand engraving can be modified with etching to darken specific areas, while an etched pattern can be detailed with hand engraving. Satin finishing and plating can further disguise the origins of the decorative patterns.

Mass-produced guns from high-volume manufacturers tend to use economical methods of decoration, where they can offer a range of different pat-

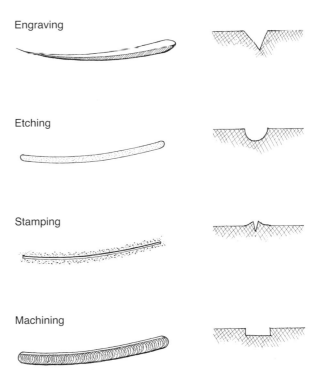

Engraving

Etching

Stamping

Machining

The effects on steel of the various methods of producing decorative patterns on guns.

139

terns for differing grades of gun. These mass-produced engravings represent very good value for money and are usually visually pleasing, even with the simplified methods of decoration used.

How can 'proper' engraving be recognized? Initially there should be evidence of good design, with the decoration laid out neatly in its allotted space. This is applied to engraved patterns, a game scene, or a combination of both. It should display the evidence of hand work in its execution, and there should be good contrast and a definite crispness to the work. It should be apparent that care has been taken even on the smallest parts of the gun, such as the forend catch.

A small amount of high-quality engraving will always be more valuable than a full coverage of factory impressed work, or rough and careless hand work. The better or more expensive the gun, the better the engraving should be; and vice versa. As in all things in life, you usually only get what you pay for.

Engraved Chequering

'Chequering' implies diamond patterns cut into the stock, but it may also be applied by the engraver to those parts that require the same non-slip function, such as the safety catch or opening lever.

Chequering can be cut with a chequering file that has parallel lines of teeth. It is designed so that multiple grooves can be cut with one stroke. Where a long piece of work is to be chequered, the teeth are overlapped a couple of rows into the previous cut to retain alignment. With small, non-flat parts, such as a top lever, it is difficult to maintain the file at the correct angle to give a consistent depth of cut. Similarly, any uneven pressure will produce deep grooves on one side, shallow on the other. It is often the case with a highly visible component that some touch-up work is necessary with a three-square or knife-edge needle file or engraving tool, attending to one line at a time.

With a second cut at an angle to the first, the conventional chequered diamond pattern appears. It is difficult to run such a pattern up a line and form a clean stop, so one method is to file a shallow line across the edge and, if necessary, reinforce the clean-cut visual appearance with a single engraved line parallel with the cut-out.

Simple single cuts can avoid this but do not have the visual impact of diamond chequering, or offer quite such a practical grip.

On many production guns the chequering will

have been stamped or moulded as part of the original casting and these patterns are easily identified. If chequering is to be applied on a safety catch, for example, it can be run out to the edge. It is even better, either with a safety or, say, top lever, to cut the chequering one line at a time with an engraving tool and enclose it within a decorative frame. Each cut is crisp and sharp, providing both good grip and visual pleasure.

Of course, chequering can also be dispensed with, and those parts where it would usually be found may be engraved to match the rest of the gun instead. Either method is quite satisfactory, as the shape of these parts is probably more important than the finish. However, fashion and economics apart, a properly chequered lever or safety catch fulfils a functional role and provides a nicely traditional appearance. They are an integral, albeit minor, part of the gun and such small parts, often taken for granted, contribute much to the pleasure of ownership and use.

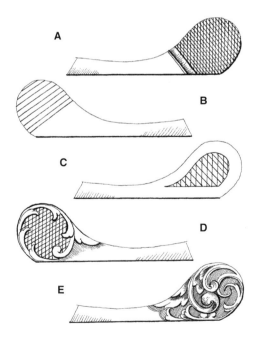

Chequering, or at least producing a 'grip' surface, done by the engraver: A is practical, effective and much better in appearance than B; design C is more often found produced as part of an investment-cast top lever; D is very effective and probably as practical as it can be, but for sheer style, design E takes the prize.

Other Uses

A secondary use of chequering or engraving is to alter a component to produce a non-reflective surface, such as on a raised top rib. Older guns may be found with what can only be described as wiggly grooves cut across the rib. These were applied, not by hand, but with a simple hand-operated 'engraving' machine.

The alternative is to use fine filed lines, most commonly single lines, at right-angles to the rib or, much less often, in diamond pattern. To cut the length of a rib by hand to produce a pattern of even depth is a job requiring considerable patience and skill, even with guides on either side of the rib to control the depth of cut. Sometimes, instead, a top rib might be matted either for part or the whole of its length. This is not regarded as engraving, but it often falls to the engraver to produce such a finish. It is a process whereby a small punch is used to hammer hundreds of tiny dents into a small area, so that any reflected light is dispersed and the surface appears matt-finished. This can sometimes be found within areas of engraving where a heavy, bold pattern has been produced to give maximum coverage for a modest amount of the engraver's time.

Inlay Work

Gold Inlay

Apart from the really exotic, such as gold roses rambling across the carved fences of a collector's piece, most inlay work is confined to lettering. One area that is very familiar is the gold 'safe' positioned just above the safety on the top tang. Indeed, most people have come to expect it to be there, and would be surprised at the amount of work that goes into producing it.

First, the lettering has to be laid out so that, when the safety catch is pushed forwards, it is completely covered, and is fully visible when the catch pulled back. Then it has to be cut out to a reasonable depth so a good thickness of gold can be inserted. Then the lettering has to be undercut to form a dovetail key – the best method – or the bottom of the cuts roughened for the same purpose. Next, gold wire is placed into the cut-out letters and hammered gently into place so the gold spreads into the undercuts, filling all the available space. Finally, the excess gold standing proud of the tang is removed with various grades of abrasive paper to leave perfectly formed gold lettering within the steel surface. The next time you take out your gun, it is worth considering that little masterpiece by your thumb.

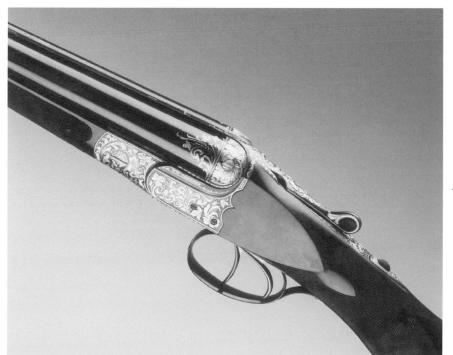

A best-quality W.W. Greener (Sporting Guns) Ltd. with gold inlay – the famous Facile Princeps model. (Photo by David Grant, courtesy of W.W. Greener)

The same principles apply for all forms of gold inlay and no adhesive or solder of any form is used, as further processes, such as blacking or hardening, would destroy them. It is the integrity of the fit, metal to metal with no air, grease or other contaminant between the two, that makes possible this sort of decorating work on guns.

Mock Inlay

Mock inlay, if done really well, can be difficult to detect with the naked eye if it is confused, say, with pale nine carat gold (gold/silver alloy). Mock inlay is produced by melting silver solder into engraving. The engraving may be cut by hand, in which case its appearance truly will be deceiving, or cut by machine, such as a hand-operated pantograph engraver, which gives a much less well-defined appearance. In spite of its description, silver solder – which is really silver brazing – is a mixture of metals of which the silver content may be as low as forty-two per cent. With added copper, zinc and cadmium, the final appearance can be much like a pale reddish-gold.

The mistake often made with pantograph-engraved letters is for the letters to be closer together than would be expected on a gun, where they are normally surprisingly widely spaced. The most obvious – and obviously bad – example of mock inlay I ever saw was produced by stamping a name with small industrial steel stamps into the side of an action, then filling with silver solder.

Commissioning Engraving

It is not only the wealthy customer who commissions engraving. The gunsmith may wish to add some embellishment to a gun, either for a customer as an improvement to an existing piece, or maybe because of the addition of side plates. Obviously there can be constraints, such as matching an existing style of engraving or working according to the customer's tastes, but, where there is a bare canvas and a free choice, it can be difficult to make a decision.

There are so many variations in design and style. Is it to be game scenes, fine or large scroll, gold-inlaid or cut lettering, deep carved fences, full or limited coverage? Colour-case hardened finish or brushed-off silver? Each engraver has his own accomplished style, and there are also well-known standard patterns that have become company hallmarks, such as Purdey's fine scroll or Holland's large 'Royal' pattern.

The best person to advise on such matters is the engraver, and most will have a portfolio of previous work that can be a starting point. Also, as the engraving will be linked very much to cost, it is best to go for a small amount of engraving of good quality than overall coverage of lesser work.

One thing to avoid is a design that may reduce the value of the gun or even make it unsaleable. Idiosyncratic game scenes involving, say, copulating animals, may attract attention but are unlikely to be to another potential purchaser's taste. It pays to be very thoughtful at this stage. Fashions may change but good taste can still be recognized thirty years later. Gun engraving, like tattoos, tends to last a very long time!

A question of taste? It certainly looks like fun, anyway.

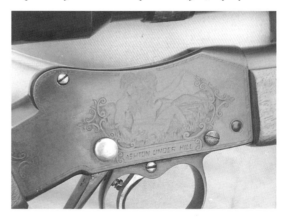

CHAPTER 13

Gun Proof

Introduction

Proof, or the proving of shotguns, is the means of testing the strength, and thereby safety, of a barrelled action by, under controlled conditions, firing a heavier load through it than the normal service load. It is a method of testing that has its critics. When any group of gunsmiths is gathered together, the topic of 'sending guns to proof' inevitably brings up the argument that the proof load may weaken a gun. Taking it a step further, some believe that it may even lead to unseen damage by bringing the barrels or action to the point of failure, and causing them to fail a short time later.

Whether this argument has any validity seems doubtful. It is true that there have been failures in the early service life of guns after passing proof, but there appears to be little real evidence to suggest that such failures relate directly to the proof test.

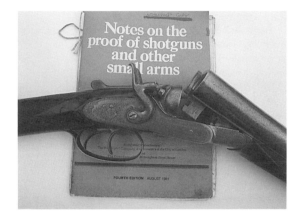

The Notes on the Proof of Shotguns and Other Small Arms, *available from Birmingham and London Proof Houses, is excellent value for money when it comes to sorting out proof markings.*

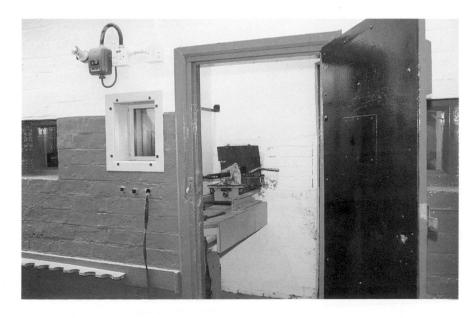

A gun (in this case, actually a rifle) in one of the Birmingham Proof House firing chambers ready for proving – after the door is shut.

That is not to say that this scenario is impossible, but it is most unlikely compared to the vast number of firearms that have been tested successfully.

On the other hand, there are a variety of reasons why guns do fail in the field – an inadvertently plugged muzzle is the most obvious, and imperfect or overloaded home loads are another. There is also the matter of 'finishing' work after proof. On rare occasions, this can be a little heavy-handed; it is disturbing to see, for example, a foreign-made magnum-proofed gun with one barrel opened up after firing less than a box of cartridges, especially when it is discovered that the barrel-wall thickness was only around 0.015in (0.38mm).

In reality the British system has worked well for over 350 years and, if military arms are included, many millions have been tested by this method without any subsequent problems. The basic method of proof may not appear technically sophisticated but, in fact, the procedures laid down by the Permanent International Commission for the Proof of Small Arms (CIP) provide a very controlled and quite sophisticated system. This, in conjunction with the experience and skill of the Proof House staff, means that the chances of an unsafe firearm getting through the system are next to impossible. It is also worth bearing in mind that a gun should be made (or designed) to be safe firing the proof load,

not made to accommodate the service load with the chance that it might go through proof!

History of Proof

It is generally believed that proving of firearms, originally muzzle loaders, was introduced because of concerns over public safety due to the number of unsafe arms being produced. W. W. Greener, in his milestone publication *The Gun and Its Development*, puts a slightly different perspective on this, suggesting that the Gunmaker's Guilds, which were keen to promote the system, may have been acting in part to curtail manufacture by rival makers who were not members of their organizations. Also, they would benefit from the existence of an independent body to prove firearms, as this relieved them of a certain amount of responsibility. Whatever the possible politics and partial self-interest behind those early moves, they led ultimately to compulsory proof of all civilian firearms, and there can be no doubt that this helped to remove 'rogue' guns from the marketplace.

Originally proving of barrels was carried out by the makers themselves or by private Proof Houses. The first move towards formal proof was in 1637, but there was little legal basis for enforcement until

The yard and frontage of the Birmingham Proof House, established 1813.

1670, when the Gunmaker's Company of London was able to enforce proof 'in London and the suburbs and within ten miles thereof'. Nearly two hundred years later, in 1813, the Birmingham Proof House was established, and from this time on it was an offence to sell, or offer for sale, any unproved firearm anywhere in the UK. Further Acts followed, in order to clamp down on suspected evasion of the 1813 Act by some makers or importers. Subsequently, laws were passed in 1815, 1855 and 1868, to further define the parameters and operation of the laws relating to gun proof.

The Act contains the Rules of Proof, which are the standardized working instructions to the Birmingham and London Proof Houses. When change to these rules is deemed necessary, new or modified rules are submitted by the Proof Houses for approval by Her Majesty's Secretary of State. This is not a frequent occurrence (nothing changes quickly in the gun trade), and the current rules were established in 1989. Earlier rules were in force in 1875, 1887, 1896, 1904, 1916, 1925, 1954 and 1986. Interestingly, any gun proved under any of these earlier rules is still legally valid for sale, provided it still conforms to its original proof size and has not been altered or weakened in any way. From a practical aspect, the gunsmith is concerned only with breech loaders that qualify as pre-1904, and mainly black-powder, and the variations in proof marks post-1904. Muzzle loaders that were under earlier rules of proof are usually sold as antiques.

The Basis of Proof

The basis of proof is the size of the bore, and the original means of sizing was arrived at by the number of spherical lead balls of a given bore diameter that would make one pound weight – a four bore would take a ball of ¼lb weight, in other words, four to the pound. Using this method, it follows that a twelve-bore would take twelve balls to the pound, and a twenty-bore, twenty balls to the pound. This system of measurement was originally not confined to smooth-bore guns but also included rifled small arms, the bore designation (what is now referred to as 'calibre') being the size prior to the addition of rifling.

Checking the bore size with either plug gauges or a comparator is known as 'gauging the bore'. 'Gauge' is the equivalent American term for the British 'bore', so a sixteen-gauge is the same as a sixteen-bore; and gauge is really the correct term.

The bore or gauge is not one exact size, but has a tolerance, or small range, of sizes within each proof size. For example, the nominal size twelve-bore covers the size range 0.729in to 0.739in (of course, it makes no sense for bore sizes originally measured under the Imperial system to be converted into metric). Once a barrel exceeds the bore size range it is deemed 'out of proof' and, as with an unproved firearm, it is illegal to sell, offer for sale, exchange or pawn it; the penalty at present is a fine of £5,000 for each offence. It is worth noting that the offence is committed by the seller, not the purchaser. Higher penalties are possible where multiple offences occur and altering or forging of proof marks is a particularly serious matter.

Obviously there are justifiable safety concerns about a gun that is in otherwise good condition but out of proof. If a customer brings in such a gun the gunsmith's first duty is to inform the customer of their legal responsibilities and strongly to recommend re-proof. What the customer does about it from that point on is their decision. The gunsmith certainly has no authority to refuse to return the gun to the owner. Indeed, the owner can continue to use an out-of-proof gun at their own risk. However, were there an accident, and a third party injured, legal complications could arise.

Many years ago, I went to our local gunsmith with another young chap who had inherited a twenty-bore double with an interesting grip safety. He proudly offered up the gun for examination and it was found to be only black-powder proof, with thin barrels, and also well out of its proof size. The young owner, to mask his disappointment, pretended nonchalance and then requested a couple of boxes of twenty-bore cartridges. The gunsmith wisely refused to sell him any cartridges and had a few choice – and quite unambiguous – words of advice regarding the intended use of an unsafe gun. It was a lesson learnt!

Black-Powder Proof

For hundreds of years, guns were tested with black powder as it was the only propellant available, but towards the end of the nineteenth century the first smokeless powders came into use. For a few more years, black powder continued as the industry standard, with proof for the later powder optional. Eventually nitro proof became the norm, but it is still possible to submit a gun for black-powder proof. If a gun was originally proved for black powder and is of a design or construction that may be too weak to accommodate the hammer blow of nitro proof, it makes sense to re-proof it with black powder. Most enthusiasts now accept that if you have an early breech loader, it is best appreciated when used with its original loadings. The mellow boom of 'old blackie' on a misty November morning, the gentle push of recoil, the smoke and smell are something to be savoured – the kind of thing memories are made of. Some years ago there was a fad for attempting to get any breech loader through nitro proof, on the basis that it had a wider appeal and, therefore, greater market value. A number of fine old guns were lost in this way, which is a shame.

Muzzle loaders, both original and reproductions, are proved with black powder, and many modern guns bear the markings 'Black powder only'. There are one or two specialist muzzle-loading pistols that are intended for smokeless powder, but they are outside the scope of this book.

Proof Sizes

Until 1986 all British proof loadings and bore sizes were Imperial, then, through to 1989, metric was introduced, but Imperial was optional. From 1989 onwards the metric system only has been used. However, the critical measurement is still taken at 9in (227mm) from the breech face and the proof size range is 0.2mm, which neatly converts to 0.008in. This means that, with the metric system, where the proof size to the nearest 0.1mm is marked on the barrels or barrel flats, it is still in proof at the second size up from the original markings when equated to the Imperial system. In other words, 18.5mm (0.728in) is still within proof at 18.7mm (0.736in). This actually makes for a more flexible system than the old British system, where 13/1 (0.728) in would have been at the extreme end of the size range 0.719in to 0.728in, becoming the next proof size at 0.729 inch.

The once universal use of plug gauges where the bore size was based on the one below the last size plug that would not enter the barrel, or at least not go as far as the 9in mark, could result in bores that were already on the 'wide end' of the proof size, as size for size does not fit and some clearance is necessary. This may offer some explanation as to why some early breech loaders are occasionally found in good condition and exhibiting little use, but are just into the next proof size when measured with a bore comparator.

Re-Proof

Re-proof becomes necessary when any of the following circumstances apply:

- when a gun is out of proof, in other words, the bores have expanded or been opened out to a larger proof size;
- there has been a lengthening of the chamber(s) to accommodate longer, and usually more powerful, cartridges, and alteration to the forcing cone(s);
- repairs such as welding have been carried out to the barrel or action;
- there is potential weakening caused by removal of metal, as in conversion from non-ejector to ejector, which requires opening out the ejector rod hole;
- either action/barrel or barrel lump have been replaced;
- any attachment such as a variable choke has been added.

Considerations for Re-Proof

In simplest terms, a barrelled action is a pressure vessel that, in use, is held in the hands a few inches in front of the face, so anything that will affect the integrity of its construction requires re-proof. When in doubt, either of the Proof Houses can be contacted and they are always willing to offer help and advice. It is also worth obtaining from them a copy of *Notes on the Proof of Shotguns and Other Small Arms*, essential reference for the gunsmith and interesting reading for any enthusiast. This also includes examples of proof marks, both UK and foreign.

Guns requiring re-proof invariably require some work to get them into a condition suitable for testing. The barrels need to have any dents and bulges removed and be struck up on the outside to a smooth surface. Any pitting in the bores must be removed as far as is practicable, and the barrels must be tight on the action.

It is normal to send a shotgun for re-proof in a stripped-down condition – without stock and forend (the barrels must be tight to the action without the forend), and, in the case of hammer guns and sidelocks, without the lockwork. Boxlocks are sent with the hammers, mainsprings and sears in place. Then it is a matter of waiting and remembering that, as my old mentor used to say, 'You can't be sure of anything until the job comes back with the marks [proof stamps] on.'

If a gun was originally proofed using smokeless

BORE SIZES: IMPERIAL		
Nominal cartridge size	**Gauge or bore as marked**	**Actual (min) bore diameter**
4-bore	$\frac{5}{2}$	1.026in
	$\frac{5}{1}$	1.001in
	5	0.976in
	$\frac{6}{2}$	0.957in
	$\frac{6}{1}$	0.938in
	6	0.919in
8-bore	$\frac{7}{2}$	0.903in
	$\frac{7}{1}$	0.888in
	7	0.873in
	$\frac{8}{2}$	0.860in
	$\frac{8}{1}$	0.847in
	8	0.835in
	$\frac{9}{2}$	0.824in
	$\frac{9}{1}$	0.813in
	9	0.803in
10-bore	$\frac{10}{2}$	0.793in
	$\frac{10}{1}$	0.784in
	10	0.775in
	$\frac{11}{1}$	0.763in
	11	0.751in
12-bore	$\frac{12}{1}$	0.740in
	12	0.729in
	$\frac{13}{1}$	0.719in
	13	0.710in
16-bore	$\frac{16}{1}$	0.669in
	16	0.662in
	$\frac{17}{1}$	0.655in
	17	0.649in
	18	0.637in
20-bore	19	0.626in
	20	0.615in
	21	0.605in
	22	0.596in
28-bore	27	0.556in
	28	0.550in
	29	0.543in
.410	Sizes for .410 are as marked on barrel	0.415in
		0.410in
		0.405in

METRIC AND IMPERIAL EQUIVALENT BORE SIZES

Metric	Imperial	Metric	Imperial
24.0	0.944	18.4	0.724
23.9	0.941	18.3	0.720
23.8	0.937	18.2	0.716
23.7	0.933	17.7	0.697
23.6	0.929	17.6	0.693
23.5	0.925	17.5	0.689
23.4	0.921	17.4	0.685
23.3	0.917	17.3	0.681
21.1	0.830	17.2	0.677
21.0	0.826	17.1	0.673
20.9	0.822	17.0	0.669
20.8	0.819	16.9	0.665
20.7	0.815	16.8	0.661
20.6	0.811	16.1	0.634
20.5	0.807	16.0	0.630
20.4	0.803	15.9	0.626
20.0	0.787	15.8	0.622
19.9	0.783	15.7	0.618
19.8	0.779	15.6	0.614
19.7	0.776	14.2	0.559
19.6	0.772	14.1	0.555
19.5	0.768	14.0	0.551
19.4	0.764	13.9	0.547
19.3	0.760	13.8	0.543
18.9	0.744	10.6	0.417
18.8	0.740	10.5	0.413
18.7	0.736	10.4	0.409
18.6	0.732	10.3	0.405
18.5	0.728	10.2	0.401

Sizes only to third decimal place as most comparators will not read smaller than this.

powder, then it is submitted for the same at re-proof. With a black-powder breech loader there is a choice, and one school of thought holds that having a gun proof-tested for modern cartridges is very much a statement of its soundness. Although this is driven in part by commercial considerations, it is also a fact that most shooters do not want the inconvenience of loading their own black-powder cartridges and the attendant cleaning ritual.

A gun that has been nitro-proofed can be used also with black powder, but the reverse is not acceptable. Obviously, early guns that qualify as collectors' items, or are rare and interesting pieces, would not normally be subject to the later type of proof. Metric proof markings are not welcomed by collectors, although miniature markings are a service provided by the Proof Houses.

Great care should be taken when selecting an early gun for re-proof. The pressures generated by nitro powders peak very quickly and put more pressure on the breech than black powder. Early breech loaders, with thin standing breeches and a beautifully crafted right-angled corner (root), where the breech face meets the action bar, are not the best candidates for nitro proof. That knife-edge corner, as it is often described, is potentially a fault line, as in the saying 'a corner is half a crack'. While it may not actually break at this point there is a good chance that the standing breech will be pushed back, producing a gap between it and the barrels, and, subsequently, a proof failure.

The locking system is another consideration. Guns with the Jones underlever are locked very strongly, but some snap actions have only a single bite, and those rare latch bolts that have a cut-out full width of the action bar, while locking up well,

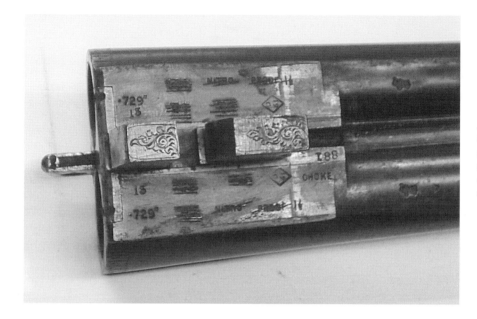

A rare sight – a pair of barrels that failed proof some years ago and were deemed irreparable, so the original proof marks were struck out.

have potentially some of the weakest actions. Bar in wood actions, where the stock work is extended around most of the action bar – with wonderfully crafted crab joints at the pivot – are not unduly strong, but good examples have been put successfully through nitro proof.

It is also important to consider the barrels. Very early breech-loading barrels were little different from those found on muzzle loaders and probably from the same source. They are identified by noticeably thin chamber walls and an elegant straight taper form, fine for black powder but not best suited to the type of pressures generated during nitro proof. However, if the breech ends of the barrels are of the later, heavier form but the barrels are thin – and 'thin' can mean as little as 0.020in (0.5mm) – prior to the chokes, nitro proof is sometimes a better bet as black powder generates more pressure further up the barrel. Where there is any doubt, it can be an advantage to open out the chokes to reduce any pressure surge. Similarly, if the barrel walls are of even thickness, or, more correctly, even taper, without any quick changes of wall thickness, they have a better chance of passing re-proof than a barrel with reasonable wall thickness but with comparatively thin areas.

The lesson here is that, if there is any doubt, a gun should not be subjected to a method of proof that it was not designed to withstand; it should be used as originally intended.

One matter that does not relate directly to proof, but is a factor to be considered with older guns, is the use of steel shot as an alternative to lead. In England and Wales lead shot has been banned for shooting over wetlands and certain sites of special scientific interest (sssi). The result is a complete ban on shooting wildfowl anywhere in England and Wales while using lead shot. There are two problems here: Damascus-barrelled guns are quite unsuited to steel shot; and, as steel is nowhere near as dense as lead, it takes up more space in a cartridge case than the equivalent weight of lead. Black powder has much more bulk than smokeless powders and, with sufficient wadding, a traditional twelve-bore 2½in (65mm) case with overshot card and rolled turnover will just hold two and three-quarter drams of powder and 1⅛oz (32g) of lead shot. Using a very good grade of black powder, such as the 'Swiss' brand, can allow a reduction of powder charge as it is somewhat more potent than many cheaper powders. However, it is still difficult to get the larger loadings by weight of steel shot into a black-powder cartridge case.

Alternatives to Proof-Testing

It is not necessary, under certain circumstances, to submit an out-of-proof gun for re-proof to enable it to be legally sold. Certain small arms that are chambered for obsolete cartridges, or are of the greatest rarity and particular historic interest, may be eligible for a 'proof exemption certificate'. This service is at the discretion of the Proof Houses, which, if the case is valid, will issue a certificate with the following statement: 'The arm is an antique or collector's piece, which is not intended to be fired, is unlikely to withstand the proof test and might be severely damaged if subjected to proof.' There is a caveat at the bottom of the certificate stating that the gun should not be offered for sale as serviceable, has not been proved and should not be fired.

As a general rule, the Proof Houses will not issue a proof exemption certificate for a gun capable of chambering modern cartridges, unless it really is a most rare and valuable collector's piece. However, they may be issued where 'work has been carried out rendering the weapon incapable of discharging any shot, bullet or other missile'. This is similar to, but not to be confused with, a deactivation certificate, and covers such guns as cutaway demonstration pieces.

Deactivation

Introduction

Deactivating is the act of altering a firearm to a particular standard so that it cannot be used to discharge any type of shot, bullet or missile. It is the most radical alternative to proof and is really just a sophisticated and relatively expensive way of scrapping off a gun. It does mean that the owner can hang it on the wall, safe in the knowledge that it cannot be used.

This is not particularly satisfying work for the gunsmith, as it is destructive rather than creative, but there is a certain demand for it, as a modest specialist market for collectors, especially of militaria, has sprung up. If a gunsmith is called upon to carry out deactivation work, it will quite likely be on an old family shotgun that is now well past its 'sell by' date. For sentimental reasons, the owner perhaps wants to keep it, but not have to bother with the hassle of obtaining a shotgun certificate. While I do not have a problem with deactivating an old gun that would be dangerous to fire and is, for practical purposes, quite beyond repair, I always refuse to deactivate something nice and useable. It is simply a matter of principle versus profit.

History of Deactivation

Prior to 1988 there was, in the UK, no formal definition of deactivation in law. Of course, there were deactivations, varying from simply having the firing pins and mainspring taken out, to having the actions welded shut, barrels blanked or holes drilled through barrel walls. Some of these were informal, and some of the more comprehensive work was done on guns that had proof exemption certificates. The trouble with such informal deactivations, no matter how comprehensive and well-meaning, was that the owner could end up in court charged with illegal possession of a firearm. Indeed, one amateur collector was so charged – he had a shotgun on the wall with its firing pins removed and a series of holes drilled through the barrel and flats for the full length covered by the forend. Actually, this old shotgun would have been of considerable risk to the user, even before the informal deactivation, if it had ever been fired!

To a great extent, this sort of confusion was cleared away with the 1988 Act, which, for all classes of small arms, laid down formal methods of deactivation for checking and certifying by a Proof House. Subsequently, for some types of small arms, the methods of deactivation have been upgraded to make any kind of attempt at reactivation by the criminal virtually impossible, or, at least, not worthwhile.

The one disadvantage of the law is its very inflexibility. By stating an exact procedure for deactivation it does not take into account ways of deactivation that may be even more comprehensive than those approved by the Secretary of State. Fully sectioned or skeletonized firearms (which were once useable guns), which logic tells us would be quite unuseable, and would, it seems, therefore qualify as a non-firearm, would not necessarily qualify in law as a deactivation. It could be argued that such a gun may sometimes still require authorization for possession.

It is worth noting that present Government proposals may outlaw deactivating and the sale of deactivations, but not ownership of existing examples.

Deactivating

There has been a tendency, when carrying out deactivations, to do the work as economically as possible, which has meant using tools such as an angle grinder fitted with a cutting disc, particularly when dealing with ex-military firearms. When deactivating shotguns made of comparatively soft steel, however, such crude equipment is not needed. It is important to make a neat job, as the gun is still the owner's property, and it is better to mill and cut away material that has to be removed in a controlled manner rather than to hack at it with a grinder.

There are subtle variations with deactivating shotguns, but the principles remain the same, and the full specifications are available from either of the Proof Houses. The most common deactivation the general gunsmith is likely to come across is that of the ubiquitous double-barrelled twelve-bore hammer gun.

The barrels, with extractors removed, should be slotted through the flats and into the barrels to within 1in (25mm) of the tip of the forend. The slot has to be a minimum of one-third of the bore diameter, so this is easy to achieve by clamping the barrels upside down in a vertical miller and cutting away with a ⅜in (10mm) end mill.

After deburring, steel plugs, tapered at one end to fit the forcing cone, can be MIG welded in place. Gas welding is not appropriate as the heat input is too great and will loosen the ribs, but it is possible, with care, to get away with electric arc welding using small-diameter rods. As far as the quality of welding goes, I have never had very satisfactory results when welding Damascus barrels to the plain steel plugs. Fortunately, when the Proof House tests

the security of the plugs, it is interested only in the strength of the joint so, even though the welding sometimes produces a version of metallic pigeon droppings, it is not too much of a disadvantage.

With the barrels finished, work can be carried out on the action. This consists of removing the firing pins, which are no longer part of the gun, and drilling out the holes with a ¼in (6mm) or ⁵⁄₁₆in (8mm) drill, and this is most easily achieved with a hand drill using the larger end of the firing-pin hole as a drill guide. Once this is done, the action can be reversed in the vice and the breech face drilled away using a series of drills, gradually stepping up in size until the last one is cartridge-rim diameter. Sometimes if the action has been case-hardened exceptionally well it will take the edge off the drill bits. Usually, though, once the first drill bit has gone through, because subsequent bits are only catching the edge of the hardening, it is not a problem. Using tungsten carbide-tipped drills for the pilot hole will overcome the case-hardening, as will localized annealing with a gas welding torch. Only on one occasion have I had to anneal the bare

Part of the Proof House laboratory, showing the present-day sophistication of its procedures. In the foreground, a test gun linked to computer. Beyond it is the jig for accurately drilling cartridges for pressure take-off, and on the wall, a variety of test barrels.

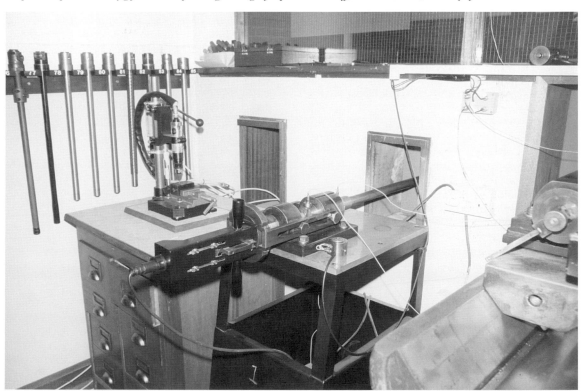

action for several hours in a furnace, to overcome the hardest piece of steel on an old action that I have ever encountered.

When the job is completed, to be accepted as a formal deactivation, barrels and action have to be submitted to the Proof House for examination, stamping and certification.

Proof-House Procedures

An item sent for re-proof needs to be labelled and identified with the name and address of the individual or company submitting the gun, the maker's name or identification, and the nominal bore or gauge size and chamber length. Many old guns may not even carry a name. In this instance, it is permissible to use the original proof marks to denote the origin of the gun, so a sixteen-bore gun bearing Birmingham proof marks could be described as, say, 'Birmingham-made, double-barrel side-by-side hammer gun 16-bore × 2½in (65mm)' (the latter referring to chamber length).

Some old guns can be found without serial numbers and this is not a legal requirement when offering for sale. However, the CIP requires all firearms for proof to be uniquely identified, and in the very near future, if an old gun is submitted for re-proof without a serial number, it will be a reason for rejection – until a serial number is added!

Once into the system, the gun is checked for cleanliness around the breech and visually examined for any imperfections, such as being off the face, barrel pitting or other damage that might result in rejection on viewing. The basic functioning is also checked to ensure that it is practical to submit it to testing.

Assuming it accepts a currently available cartridge, the proof-test load is thirty per cent greater than the maximum pressure generated with a service load. In Proof House terms, this is P T max (pressure transducer maximum), derived from the method of testing. For example, a twelve-bore 2¾in (70mm), in other words, 12-70 with a P T max of 740 bar will be proofed with a minimum proof load at 960 bar. Not only that, but the proof-charged ammunition is tested in pressure test barrels for consistency in batches of ten. Readouts are via pressure transducers linked to a computer.

After proof firing, the gun is visually checked again for any faults such as gaps that may have appeared at the breech, and bulges or rivelling in the barrel(s). If a gap is visible anywhere between the breech and barrels, it is checked with a 0.002in (0.05mm) feeler gauge. If this enters the gap at any point the gun is reject after proof. Examination of cartridge cases after firing may also point to deep-cut rim recesses (giving excess headspace), or movement between barrels and action during firing.

The Proof Houses have written procedures to cover all their methods, with flow charts to show all eventualities, such as obsolete calibre, proof ammunition not available, pressure test barrel not available in a particular calibre, and so on, and the appropriate action to be taken in each case. This is leading up to ISO approval, so standards and procedures in all countries signed up to CIP should be the same.

CHAPTER 14

The Law

Introduction

The UK has some of the most restrictive and complex legislation in the world governing the civilian ownership and use of firearms; sometimes it even seems rather illogical. There are a number of reasons for this, but, clearly, it is a fact that many in Government regard the civilian ownership of firearms of any type as undesirable. Any means to restrict access to, or ownership of, firearms is regarded as beneficial, whatever the truth of the circumstances. Major revisions have been made to existing legislation as a political reaction to various firearms-related incidents or tragedies. Most of these political gestures have simply imposed more restrictions on legitimate shooters, given more power to the licensing authority (the various police forces), and done little to deal with the system failures and social problems that led to the tragedy in the first place.

A number of pressure groups, assisted by some high-profile figures from the acting world, are now adding their demands to central Government. Most, it would seem, have little real knowledge of the various facets of sporting shooting, but they continue to air their prejudices in public. Most, if not all, shooters do accept the need for controls, framed to protect society while maintaining the rights and freedoms of those who choose to possess and use firearms responsibly. By contrast, the various 'gun control' groups, who are actually tiny in number and emotive in argument, have only one ultimate aim: the total prohibition of civilian ownership of firearms.

In 2005, the new Home Secretary announced his intention to ban the sale, manufacture and import of 'realistic replica' firearms, most of which are little more than toys. Eighteen months ago the same Government enacted a law that effectively prohibits children playing cowboys and Indians in a public place. This was at a time when real crime

with pistols, most of which were banned in 1997, and automatic weapons (banned 1936), is increasing at an unprecedented rate in modern times. (Incidentally, 'ban' really means re-categorizing. These items do not disappear but require Home Office approval for legal possession, which is very difficult to justify.)

It is not possible to do the complexity of UK law full justice here, so this chapter will provide some background to the present day, and a guide as to how the law affects the aspiring gunsmith. It is, however, very much an overview. Real detail is found within the Firearms Act, but that can sometimes take considerable interpretation and the best answer is to join a relevant organization, such as the Gun Trade Association, which is well versed in the law and has access to some of the most experienced experts on the subject.

History

The main act in this process of gun control dates back to 1920. The British Establishment was reeling from the fall of the Russian Tsar, following the Bolshevik Revolution, and concerned that similar events could overtake Britain. Suddenly, tens of thousands of 'Tommies' back from the trenches were no longer to be trusted. Many believe that, had there not been a war in 1914, there would have been enormous pressure for social change in these times. But, post-war, people were too tired of conflict, and too many sons and fathers had never returned, or, if they had, it was often with their health in ruins. History also showed that Britain as a nation was disinclined towards violent revolution.

Social uprising was, therefore, unlikely, but the mass of the people were none the less seen as a possible threat to the status quo. As power comes from the barrel of a gun in the early stages of revolution, civilian gun ownership had to be restricted. It was at this point that the deceit began – none of these fears was aired publicly in Parliament. Instead, the alleged need for firearms controls was wrapped in a web of mistruths. The new law was necessary, it seemed, to keep firearms out of the hands of criminals and lunatics. It was all for the good of the British public at large.

It was the beginning of an established pattern of law-making, continuing over some eighty-five years, in which attempts were made to link legitimate firearms ownership with armed crime and, later, the misuse of a legitimately held firearm by an individual would result in punitive laws against all other gun owners. Finally, Government has given up on the myth of legitimate ownership linked to crime as it has been unsustainable following the rise in pistol crime after the 1997 ban. Now a new chapter has opened: gun-control laws are, we are told, for public safety.

Whatever the origins, lack of logic and injustices of the Firearms Act, it is there to be obeyed and the vast majority of shooters are law-abiding people. There are few worse examples of bad law and no finer example of adherence to such a law than the pistol ban (actually a ban on firearms below a certain length). Ultimately some 80,000 shooters and hundreds of registered firearms dealers had to hand in their prized possessions. They had fought with the pen and by peaceful demonstration, by speeches, marches and appeals to their parliamentary representatives – all to no avail. When the time came, they handed in their firearms with quiet dignity.

Shotguns

Most restrictions have concerned rifled firearms. There were few controls on shotguns prior to the Criminal Justice Act of 1967, when only a ten-bob (50p) annual licence was required from the Post Office (a revenue-raising measure dating from 1873) to possess a shotgun beyond the confines of the home. No licence was necessary to purchase or to keep a shotgun at home.

In rural areas, shotguns were kept in sheds and barns, wherever they were handy for use, and, in spite of sawn-off shotguns occasionally being the choice weapon of urban bank robbers, there were no real restrictions (although law breakers were, of course, punished if caught). One of the results of the 1967 Act, which followed the tragic shooting of three police officers by criminals using an illegally held revolver, was to impose a shotgun licensing system. Under this system, the individual was

licensed but there was no restriction on the number of guns that could be held. In 1973 a soon-to-become-infamous Green Paper recommended sweeping changes to the Firearms Act; it was considered by, and convincingly thrown out of, Parliament although it continued to be quoted as a reference by various anti-gun groups for the following fifteen years.

The next big impact on shotgun ownership followed the shootings in Hungerford by Michael Ryan, who armed himself with rifle and pistol. For the 1988 Amendment Act, many of the proposals of the 1973 Green Paper were dusted off and subsequently incorporated, albeit sometimes in modified form, into law. These included a licensing system that recorded individual shotguns possessed, magazine restrictions on pump-action and semi-automatic shotguns, security requirements and limitations on loaning a gun. Interestingly, by this time individual awareness of adequate security was much better than in previous years and the use of shotguns in crime was in decline.

The 1997 Amendment Act was not really directed at shotgun ownership, and the situation still exists that the individual has certain rights to possess a shotgun, unless he is debarred by previous criminal convictions or the Chief Officer of Police can prove he is an unsuitable person. Neither is there a limit on the number of shotguns that an individual can possess, except for those of suitable security. Needless to say, various anti-gun pressure groups would like to see this changed, to oblige the applicant to prove good reason for possession of each or any shotgun, and to limit the numbers that may be held; this is in spite of shotgun crime being at an all-time low.

Registration as a Firearms Dealer

Process of Registration
To be a gunsmith dealing in most types of repairs it is necessary to register with the local constabulary as a registered firearms dealer (commonly referred to as RFD). Prior to acceptance the Chief Officer of Police will wish to be satisfied that:

- the applicant is not prohibited from possessing firearms by virtue of previous criminal convictions or issues of mental health;
- the applicant can be permitted to carry on a firearms business without danger to public safety or the peace; and
- the applicant will engage in business to a substan-

tial extent or as an essential part of another trade, business or profession.

In other words, a dealer's licence is not available to an applicant who simply wants to build up a collection of firearms.

As part of the checks, the Chief Officer of Police (in real terms, the firearms department) will consider the character, antecedents and background of the applicant, and his experience and knowledge of firearms and the Firearms Act, and of related matters, such as the Proof Laws.

The process of evaluation is helped if the applicant is already a shotgun or firearms certificate holder, as they will already have been interviewed by the police, and will therefore be 'known' to the police (for the best reasons). This does not, however, debar the applicant who is not a shooter – and, interestingly, many workers in the gun trade have not been active shooters. For example, an applicant who wants to register as a firearms dealer may have had experience of firearms whilst in military service, while practical shooting experience would not necessarily be an advantage for someone who wishes to set up as an engraver. One stocker I know has hardly ever fired a gun except when it has been beneficial for testing.

Some local authorities (councils) request that the police inform them when an individual applies for registration as a firearms dealer. However, passing on such information may qualify as a breach of confidence. Their argument is that the applicant may be in rented local authority housing, which usually prohibits running a business from home. (It may also be to do with potential loss of revenue.) In any case, the individual must sort out any possible planning matters on basic principles at a very early stage, perhaps even prior to the application. Planning permission is not necessarily essential as permission may be gained to operate a business from a private workshop, but only for the applicant, with any commercial use coming to an end when that business ceases. It is still necessary to pay commercial rates, usually at a lower rate than a full commercial premises.

Once all these hurdles have been overcome, the application form filled in, the fee paid and all interviews with the visiting firearms officer satisfactorily completed, registration as a firearms dealer will become a reality. The local police authority will add the applicant to the register of firearms dealers in their area and the applicant will receive a certificate of registration as a firearms dealer, which lasts for three years, and bears a unique reference number that is quoted in all transactions.

Security is obviously a major issue when it comes to storing quantities of firearms, even though many may not be in full working order. The place of business is an essential part of the registration and security must be satisfactory for the type and volume of estimated business. Sometimes on the certificate of registration limits are placed on the number of firearms to be stored at any one time, if the security or storage space is limited (*see* page 156, 'Security'). If there is any dispute on this point, the *Home Office Firearms Security Handbook* is available on the Home Office website.

The registration as a firearms dealer is specific to the registered premises or place of business. If a registered dealer has more than one place of business, they all have to be registered. Should any fall within another police force area, they have to be registered with the other constabulary. Operating at another venue such as a game fair also requires a temporary registration for other places of business with the appropriate police authority.

It is important at the time of application to be accurate as to the scope of the intended business as this will be reflected on the certificate of registration. For example, 'Repair of shotguns' excludes any involvement with Section One firearms (rifles and certain types of pistol) and shotguns that qualify as Section One firearms. Normally once a dealer is registered, it covers all Section One firearms and Section Two shotguns. There are not normally any restrictions.

Administration of the Firearms Act
As this book goes to press in January 2006 there are forty-three separate constabularies in England and Wales and eight in Scotland, which administer the UK Firearms Act, although this may be subject to a reduction. The Isle of Man has laws that are generally based on the UK but modified by the Manx Government to suit its own requirements, and Northern Ireland also has its own firearms Act. When dealing with customers from either of these areas it is best to contact the relevant constabulary if there is any doubt about compliance with the law.

The existence of so many independent UK constabularies is to allow for different circumstances – a large, mainly urban police force will face different problems from, say, a mainly rural force. The real effect sometimes is for senior officers to impose their own interpretation of the requirements of the Firearms Act so that they may be subtly different from a neighbouring area, even one that has similar policing requirements. The substantial Home Office document *Firearms Law – Guidance to the*

Police (publicly available) is intended to promote the same interpretation of the Firearms Act by all police forces. In the main this seems to work, and many constabularies apply the law even-handedly, but there are occasional problems where individual officers try to 'improve' the law. Most minor perceived problems or misunderstandings can be overcome by face-to-face discussion. A new applicant may believe himself to be in the right on a point of law but, if he comes up against a member of the firearms department staff who has decided to 'improve' on the law, quiet, but factual and firm diplomacy is probably the best approach.

The Register

One of the most important requirements of a gunsmith is to keep a register of all transactions. There are commercial registers available, but most constabularies will accept case-bound lined notebooks of A4 size. According to the Home Office guidance to the police, ring-bound books or simple card index systems are not acceptable.

Entries are normally laid out with acquisitions or work received on the left-hand page, and sales or work out on the right-hand page. The left-hand page may be laid out in the following manner:

Date	Name, address and certificate number of owner	Quantity	Description including serial number of gun (if applicable)

In addition it can be useful to include a column for the work to be done or the reason why the item was received in.

On the right-hand page, the return of a repaired shotgun is recorded with the date returned and details of the person to whom it has been returned. With shotguns the customer should show their certificate at the time of bringing the gun in for repair. There is an exemption for persons collecting shotguns from a gunsmith after repair, test or proof. However, it is not unreasonable for the gunsmith to ask to see a valid certificate on collection.

It can be convenient and it is quite acceptable to sub-divide the register, or to keep separate registers for repair work and those items held as stock or perhaps even manufactured. This becomes a much more comprehensive exercise when dealing with rifled or Section One firearms where major component parts, ammunition and expanding bullets all have to be recorded.

Following a 1997 amendment to the 1967 Act, it is acceptable to keep the register details on computer. However, it is vitally important to keep a back-up of data as a safeguard in the form of floppy disks or other media, as well as hard copies of transactions, dated and page numbered.

Current legislation requires the register to be kept for five years from the date of the last transaction.

Security

Security for commercial premises always used to be the province of the local crime prevention officer, but in some cases this has now been absorbed as one of the functions of the relevant police firearms department. There are a number of factors to be taken into account.

The basic security of the building, and its location and distance from inhabited property all need to be taken into account. A workshop some little distance from the house on the same piece of land obviously has some inherent security advantages over a shop in town that has no authorized human presence overnight. Similarly, in a workshop it is not at all difficult to build a store from concrete blocks with steel reinforcing and a concrete roof. Add a fire door clad in steel, hinge bolts and two good seven-lever locks, and there is the basis of a secure store. With a shop, many items are required to be on display, obviously as an aid to sales. There may not be room for a workshop-type store, or the premises may be leased and the landlord less than enthusiastic about any alterations. The other weakness with a shop is its most potent sales device, a large window, and also the ceiling, where access may be forced via the roof and upstairs room. Most shops have a comprehensive intruder alarm system and there are plenty of security or alarm companies willing to offer advice; some are much more expensive than others. It is important not to forget to involve the firearms officer; he or she is the person with whom, ultimately, agreement has to be reached regarding the level of security. Some regard the firearms officer as a nuisance, getting in the way of the gunsmith making his living, but most of them are actually very helpful and more concerned with reaching agreement on an adequate level of security than impeding the business.

Once the security arrangements have been agreed, which will be a relief to any insurers, unless there is a change in the nature of the business the police cannot later legally demand to have them upgraded. However, if the incident rate of crime such as burglary is on the rise locally, it may be worth reviewing security; the crime prevention officer will probably heave a sigh of relief.

Where the amount of business is limited – perhaps because it is part of another business – and either a small number of guns or only parts are kept, security can be tailored accordingly. A good example of this is the engraver or colour case hardener, who will only have certain parts, some of which may be unrelated and, anyway, without any lockwork or internals. In such cases a steel cabinet secured to brickwork may be acceptable. These are not usually the smaller commercial gun cabinets, but either purpose-made or even a commercial safe.

In certain instances, for someone who tests or works only on component parts of shotguns, it is not necessary to be registered as a firearms dealer. For example, someone setting up as a blacker, who would only have the barrels, furniture or perhaps a stripped-down bare action in their workshop, could be exempt from the requirement to be licensed. It it worth discussing with the firearms department if a business is to be limited to specific functions, and completed shotguns will not be handled at any time. The same exemption does not apply to rifles or pistols where the major component parts – barrel, bolt or block and receiver – are licensable items.

The Home Office guidance to the police 16.21 summarizes the security situation as follows:

> The conditions are intended to ensure that a registered firearms dealer takes reasonable precautions for the safekeeping of their firearms and ammunition. Crime prevention officers, in consultation with their firearms departments, should consider carefully the level of security required in the light of the circumstances of each case. Circumstances may vary between one district and another and between one dealer or premises and another. Much will depend on the degree of risk and the steps that it is practicable to take.

Police Inspections

Although the certificate of registration lasts for three years, dealers' premises will often be visited annually and, indeed, inspections can be carried out at any time by either police officers or civilian support staff authorized in writing by the Chief Officer of Police. Such written authorization must be produced if requested, although the firearms department staff do become familiar faces. Often the frequency of visits will depend upon the size and scope of business; checks being limited to record inspections or, in the case of renewal at the end of the three years of the registration, both stock checks and records inspection. Total numbers of transactions may be recorded so the Chief Officer of Police can be satisfied that the trade is consistent with the stated intentions of the dealer.

The kind of information that is audited is as follows:

- sale or transfer of a firearm to a certificate holder within the police force area;
- sale or transfer of a firearm to another registered firearms dealer;
- sales or transfers to a certificate holder or other registered firearms dealer in another police force area.

Where the bulk of the work is repairs, this might be interpreted as work transferred between registered firearms dealers within the same area or another area.

When an audit is completed satisfactorily, the inspecting officer will sign off the book. The Home Office advises that all open entries – uncompleted transactions where guns are in stock or not returned to customers – are then brought forward. This considerably simplifies the next inspection as it saves having old registers with one or two open entries left in them.

Each year a stock audit, including work in the 'day book', should be carried out and at this time it is useful to transfer all existing items forward into the next year's register, allocating a separate register for each year.

HM Customs & Excise

The modern-day equivalent of the revenue man feared by eighteenth-century smugglers may also audit your stock register records and, realistically, anything else relating to the business. This is to counter or deter illicit imports and diverted exports; something which will not normally involve the ordinary gunsmith serving the local community. When visits do take place it is recommended that they do so in conjunction with the inspecting officer from the local constabulary, to minimize disruption to legitimate trade.

Sale of Shotguns and Cartrides

In order to purchase a shotgun, a customer must have a valid shotgun certificate issued by their constabulary and must be over seventeen years of age. (Under the present Government, this age limit may be raised to eighteen.) In the case of a person under seventeen there will be restrictions added to the shotgun certificate, which will clearly indicate that the holder may not purchase guns or ammunition. In other words, the certificate should be read carefully and nothing should be taken at face value.

In the case of the sale or let on hire, loan or gift of a shotgun, the details of the purchaser have to be entered in the right-hand page of the register. The purchaser's title, name, address, shotgun certificate number and the date of the transaction should also be recorded.

Notification must be sent to the customer's Chief Officer of Police in writing within seven days of the transaction. This means sending by Recorded Delivery and keeping on file a copy, plus the stub of the Recorded Delivery slip, so that it can be cross referenced to the main register. Although the law states that Recorded or Special Delivery must be used, some constabularies now accept e-mail or fax notifications, but in any event hard copies should always be kept as back-up records. In addition to the recording in the register and notification to the Chief Officer of Police, the transaction has to be recorded on the appropriate section of the purchaser's shotgun certificate. On the odd occasion when the sale of a shotgun involves a foreign national, notification must be made to the police within forty-eight hours including, if available, the visitor's passport details.

No record of sales transactions of shotgun cartridges is required, but a purchaser must produce a valid shotgun certificate. For purposes of certification a shotgun cartridge must contain, or purport to contain, at least six shot, none of which exceeds .36in (nominally 9mm) diameter, this being the old British LG shot size. Cartridges containing less than six shot or sizes larger than LG can only be sold to someone possessing a valid firearms certificate with authorization to purchase.

To sell only shotgun ammunition, say as an ironmonger or farmer's supplies company, it is not necessary to register as a firearms dealer, although a valid shotgun certificate is necessary to purchase the original stock.

Mail Order Sales

Following an incident when a criminal stole a certificate and then obtained a gun by mail order, receiving it at an address different from that on the certificate – seemingly a one-off act – the law on mail order was changed. Rather than insisting that purchase by mail order had to be delivered to the address on the certificate, new law required the transfer to take place face to face. In real terms this means that a certificate holder must make an arrangement with his local gunsmith or gun shop to receive the gun from the vendor and the actual transfer takes place between the purchaser and his local dealer.

However, in the event that a gun is being returned to its rightful owner, who is in possession of a current shotgun certificate (and appearing listed on that certificate), the gun may be returned by carrier. Certainly, if some time has elapsed since the start of the job, it is worth receiving proof that the owner still has a valid certificate. For most repairs the owner will produce their shotgun certificate on bringing in the gun to the workshop, and again when picking it up after repair.

Guns Handed In

Sometimes approaches are made to registered firearms dealers by individuals wishing to hand in a gun. Often the gun has been found in the attic of a relative during a house clearance. Occasionally it is an old gun that was not registered following the 1988 Amendment Act. Prior to that, although shotgun certificates had been a requirement for twenty years, they did not list individual guns, so some individuals, on giving up shooting with the imposition of the 1988 Amendment Act, simply left them in the attic or the back of a cupboard.

Most constabularies took the pragmatic and positive view that a gun handed in in this way was another unknown now on record and away from criminal hands. Some individuals who handed the guns in to dealers were already certificate holders in their own right and would be quite happy to have recorded in the register, for example, 'Received from … , one … formerly property of …, now deceased'. Others, while doing their law-abiding duty or maybe not having a certificate, would prefer not to declare their name. Occasionally a complete stranger might arrive with a couple of guns and simply deposit them at the door, in which case the entry in the register can be no more than 'person unknown' and the details of the gun. The Home Office recommends to the police that in such an event dealers should be positively encouraged to notify the police, although they give no reason for this approach and it is a recognized fact that criminals do not normally hand in their illegally held guns.

However, there are benefits in first checking with the Gun Trade Association, which has access to the police national computer register of stolen firearms. Very occasionally, the gun will be found to have been stolen at one time, and then it is clearly a duty to notify the police.

Since the relatively recent introduction of a mandatory five-year prison sentence for illegal possession of small firearms (pistols), people are disinclined to hand in 'found' guns, either to the police or to a dealer for fear of prosecution; as far as they know, illegal possession means any gun. Like much blanket legislation, where the offence is absolute and the moral innocence of the individual is not taken into account, it has actually proven in some circumstances to be counter-productive. The problem from the gunsmith's point of view is that, once a gun has been handed over, it is not legal to refuse to accept it by handing it back to the individual who brought it in.

Scrapping or Destruction of Guns
When a gun no longer serves a suitable or safe purpose the decision may be taken to scrap it off, and this will include a gun handed in by a customer for destruction. It is useful to strip a factory gun for any salvageable parts and, as most gunsmiths cannot bear to see anything go to waste, old hand-built guns are also stripped, even to the extent of taking off the ribs. All the pins, springs and bits and pieces, which are unique on an old hand-built gun, end up in boxes, in the hope that they may prove to be useful one day.

The problem is that the parts left over still look like a gun, so they have to be rendered inert usually by cutting up, which costs time and money. Another method, if the local constabulary finds it acceptable, is to list the residue of a scrapping-off session and hand it into the police for destruction, requesting a signature for the parts. A registered firearms dealer can scrap off whatever they choose but this must be recorded in the register, and it is advisable to record also the method of destruction.

Servants
A registered firearms dealer may find it useful, from time to time, to use a servant for delivery or collection of guns and ammunition. In the Firearms Act there is no definition of what constitutes a 'servant' – a paid employee would certainly fall into the category, but it is also generally accepted that a servant may be part-time and not receive direct payment for their services.

Some police forces ask to be notified of a proposed dealer's servant and, while this is not strictly a requirement in law, it can save a lot of problems; for example, a neighbouring force may not like the idea of part-time servants. With the agreement of your local constabulary, you have the backing you need in the case of a dispute. If that servant is a shotgun certificate or firearm certificate holder, so much the better, as it makes any police checks much easier. If an individual is acceptable to the local police a simple contract should be drawn up between the dealer and 'servant' and, when on business, a letter of authorization and copy of the dealer's certificate of registration – suitably overwritten 'copy only' and signed – should always be carried.

In law, any person carrying firearms to or from a Proof House is actually exempt from the provisions of the Firearms Act as long as the items are being carried for the purposes of proof.

Out-of-Proof Guns
It is illegal to sell an out-of-proof gun, even to a registered firearms dealer. This seems to make little sense in law, as there is no advantage in terms of general safety. Guns used to be sold at auction out of proof to registered firearms dealers only, but for some years now, after careful scrutiny of the Proof law, such guns are sold as stock and actions with vee-shaped cut-outs in the barrel(s), so the only option is to rebarrel or sleeve. There is little doubt that some good guns, where the original barrels could have been saved, have been lost in this way.

An out-of-proof gun can be given to a dealer for repairs and for submission for re-proof. If the gun fails re-proof it is still, say, in the case of a private individual, that owner's property and, while the gunsmith may wish to put in writing the condition of the gun, it must be handed back to the owner if requested.

Definition of a Shotgun in British Law

This book is about shotguns, so the previous overview of the firearms laws relates specifically to them and does not include direct reference to other firearms that are more strictly controlled. Neither is it a fully comprehensive statement of the various laws as, at the moment, laws are being enacted with almost indecent haste. Much of the new legislation is ill conceived, even unnecessary, and sometimes contradicts existing established law. At the moment, one of the definitions remaining unchanged is that of the shotgun, which may, in some circumstances, be also the more restricted Section One firearm. A shotgun is defined in law as the following:

A smooth-bore gun (not being an airgun), which

- has a barrel length not less than 24in in length and does not have a barrel with a bore exceeding 2in in diameter. In law the length of barrel is measured from the muzzle to the point of ignition (breech face). For a muzzle-loading gun the point of ignition may be taken as the touch hole or nipple that is nearest to the breech;
- either has no magazine or has a non-detachable magazine incapable of holding more than two cartridges (as approved by the Proof House if not as original);
- is not a revolver gun.

This means that a shotgun with a barrel of, say, 20in – which used to be the old legal minimum – is a firearm. However, it does not, even temporarily, become a firearm if the barrels are cut back to facilitate barrel sleeving.

Conversions to .410 of military rifles, such as the Lee Enfield, will not feed from the magazine. This was recognized at the committee stage of the 1988 Amendment Act, but they were still included in the requirement to have fixed magazines reduced to at least a two-shot capacity.

Shotguns that retain a magazine capacity of greater than two cartridges become Section One firearms in law, just the same as rifles. They are no different to work on than Section Two shotguns, but the owner does need to have the appropriate firearms certificate. The exception is guns with restricted magazines for magnum cartridges – as tested by the Proof House – which will sometimes accept three much shorter cartridges. Whether they will always function properly is a different matter.

There is no legal maximum length for shotgun barrels so it is quite acceptable to have a punt gun with a 6ft (2m) barrel, or even longer, as long as the bore diameter does not exceed 2in. On that basis a small cannon with a 24in barrel and 2in bore qualifies as a shotgun.

The 1988 Act did raise to prohibited category (with the exception of .22 rimfire) all self-loading or pump-action shotguns with barrels of less than 24in or an overall length of less than 40in, excluding detachable or retractable butt stocks.

The Firearms Act applies to all shotguns, but muzzle-loading antiques are exempt from these legal restrictions and can be possessed without a certificate unless they are to be used, in which case they come under the relevant parts of the Firearms Act. Modern-made reproductions, which may be exact copies of originals, still require a shotgun certificate or firearms certificate as appropriate, even if they are not intended for use. This does mean, though, that in law a reproduction Brown Bess musket with a bore size of .740in and barrel length of 39in is a shotgun and may be used as such. Sounds a good idea for the next farmer's day clay shoot.

Useful Addresses

Birmingham Proof House
Banbury Road
Birmingham, West Midlands
Tel: 0212 643 3860
Fax: 0121 643 7872
E-mail: info@gunproof.com
Web: www.gunproof.com

Birmingham Science Museum – Thinktank
Millennium Point, Curzon Street
Birmingham, B4 7XG
Tel: 0121 202 2280
Web: www.thinktank.ac

British Shooting Sports Council
P.O. Box 11, Bexhill on Sea, TN40 1ZZ
Tel/fax: 01424 217031
Web: www.bssc.org.uk

Crown Prosecution Service
50 Ludgate Hill
London, EC4M 7EX
Tel: 020 7796 8000
Fax: 020 7796 8650
E-mail: enquiries@cps.gov.uk

Department of Trade & Industry
Export Control Organisation
Kingsgate House
56–74 Victoria Street
London, SW1E 6SW
Tel: 020 7215 8070
Fax: 020 7215 0521
Web: www.dti.gov.uk/export.control

Department of Trade & Industry
Import Licensing Branch
Queensway House, West Precinct
Billingham, Cleveland, TS23 2NF
Tel: 01642 364 333
Fax: 01642 364 269
E-mail: enquiries.ilb@imlb.dti.gov.uk

Gun Trade Association
P.O. Box 43, Tewkesbury
Glos, GL20 5ZE
Tel: 01684 291868
Fax: 01684 291864
E-mail: enquiries@guntradeassociation.com
Web: www.guntradeassociation.com

H.M. Customs & Excise
New King's Beam House
22 Upper Road
London SE1 9PI
Tel: 020 7865 5808
Fax: 020 7865 4961
Web: www.hmce.gov.uk

Health & Safety Executive
Explosives Division
Magdalen House, Stanley Precinct
Bootle
Merseyside, L20 3QZ
Tel: 0151 951 4025

Home Office
50 Queen Anne's Gate
London, SW1H 9AT
Tel: 020 7273 2184
Fax: 020 7273 4028
Web: www.homeoffice.gov.uk

London Proof House
48 Commercial Road
London, E1 1LP
Tel: 020 7481 2695
Fax: 020 7480 5102

Royal Armouries Museum -- Leeds
Armouries Drive
Leeds, LS10 1LT
Tel: 08700 344344 (24-hour info)/0113 220 1916
(info desk)
Web: www.royalarmouries.org/leeds

Further Reading

Some of the books listed were first printed many years ago, some have been reprinted.

Akehurst, Richard, *Game Guns and Rifles*, G. Bell & Sons Ltd. 1969

Angier, R.H., *Firearms Bluing and Browning*, Stackpole Books (Arms & Armour Press 1936)

Boothroyd, Geoffrey, *Shotguns and Gunsmiths – The Vintage Years*, A & C Black 1986

Brown, N., *British Gunmakers Volume One -- London*, Quiller Publishing Ltd. 2004

Brown, N., *British Gunmakers Volume Two -- Birmingham, Edinburgh and the Regions*, Quiller Publishing Ltd. 2005

Crawford, J.A. and Whatey, P.G., *The History of W & C Scott, Gunmakers*, Second Edition, Roland Ward's 1985

Crudgington, I.M. & Baker, D.J., *The British Shotgun – Volume One 1850–1870*, Barrie & Jenkins 1979; *Volume Two 1871–1890*, Ashford 1989

Greener, W.W., *The Gun and Its Development* (first published 1881, reprinted many times)

Greener, William, *The Gun 1834*, Normount 1971 (first published 1834)

Mills and Barnes, *Amateur Gunsmithing*, The Boydell Press 1986

Sparey, L.H., *The Amateur's Lathe*, George Newnes Ltd 1948

Teasdale-Buckell, G.T., *Experts on Guns and Shooting*, first published Sampson, Low, Marston & Co. Ltd. 1900

Glossary

Introduction

The gunsmith's language uses both ancient and modern descriptive terms, most of the latter deriving from factory-produced guns not of British origin. However, even the older, traditional names were not always formalized or written down, consequently they have become very much a dialect-based language with the regional differences which that implies. Many parts do have common names, while there are others that are described by subtly different and occasionally completely different terms. It would be easy to make an apparently authoritative and sweeping declaration of what should be the correct terms, but this would be to ignore the rich regional variety, which is an important part of our heritage. Neither do I believe it appropriate for a mere writer and gunsmith to decide what ought to be the acceptable standard, so I have striven to include most of the descriptive variations where they occur. This includes the traditional and differing descriptions and modern part names for side-by-sides, over-and-unders and singles, also 'modern' pump actions and semi-automatics and terms for processes, many peculiar to the gunsmith's art.

For clarity, in Section One, where there is more than one traditional name, these appear in the heading. Where modern terms appear or there is doubt as to the degree of use of an older descriptive word, it appears in the text.

Section One

Singles / Side-by-sides / Over-and-Unders

Action That part to which the barrels fit and which, with a boxlock, holds the lock mechanism.

Action bar/Bar The forward projection of a conventional double-barrelled side-by-side action from the standing breech to the knuckle.
Action body/Body Although it is the dimension of the action bar that governs the length of the action, it is acceptable to describe a gun as having a short or long body when differing from those of fairly standard length. Body is the older term but action is often used to describe the same thing. Effectively the body is the action without any mechanism fitted; a modern term is 'frame'. When complete it is usual then to refer to it as the action.

Action face The front face of the standing breech which the barrels close against; sometimes called a breach face.

Action flats The flat top of the action bar of a side-by-side and some singles which the mating barrel flats fit against. In the USA this is sometimes called the watertable.

Action slots The slots cut into the bar of hinged break-open guns to accommodate the lump or lumps. Also with an Anson & Deeley boxlock the slots machined to contain the lockwork. Conventional side-by-sides have front and rear action slots divided by a bridge.

Articulated trigger The front trigger of some double trigger guns has a pivoted, sprung joint allowing the front trigger to be pushed forward when the finger engages the rear trigger.

Back action A gun where the complete lockwork including the mainspring is situated in the stock to the rear of the action.

Bar action A type of gun where the side lock plate is set into the action bar. Where the mainspring fits onto the bar of the lock plate it is a bar action lock plate with bar action mainspring (usually bar action for short). Where a bar action lock plate has the mainspring at the rear it is correctly a bar action lock with back action mainspring. However, it is now more common for these to be described as back action locks.

Barrel face/Breech face/Breech end The end of the barrel(s) that closes against the action face.

Barrel flats With side-by-side and some singles, the flat section under the barrel either side of the lump(s) that sits against the action flats with the gun closed. Usually the area where the proof details are either engraved or stamped.

Barrel hook/Hook The hook is formed into the front face of the front lump (side-by-sides, singles and some over-and-unders) to engage the cross pin. Some over-and-unders have twin cut-outs either side of the barrels that fit onto short hinge discs. Sometimes referred to as bifurcated hooks.

Barrel lump(s) Very occasionally called steels, these are the projections between the barrel flats of a side-by-side that are shaped to both hinge and lock the barrels to the action. Usually there are two, the front and rear lumps.

Barrel tube(s) The barrel(s) of a shotgun may also be referred to as the tubes, although it is more common to use this term to describe the barrels prior to finishing, i.e. the component parts such as chopper lump tubes.

Bead/Foresight bead Small round brass or German silver bead with a threaded leg that screws into the top rib just behind the muzzle. Other usually larger variations are normally described as foresights rather than beads.

Bead May also be a small raised carved section around a ball fence, then described as a ball and bead fence.

Bend/Drop The measurements taken from an imaginary line across the top rib to the front of the comb and heel of the stock.

Bent/Hammer notch The cut-out in the tumbler or the internal hammer which the nose of the sear engages to hold the mechanism cocked. While bent is a provincial and Birmingham term, hammer notch tends to be the favoured description in London.

Bite(s) The cut-out in the barrel lump or lumps where the locking bolt engages. When only one bite is used it is a 'single bite' action, when two are used a 'double bite' and three – where there is also a top bite – a 'triple or treble bite'.

Bolt Device that slides in the bar of the action to engage the bites and lock a gun shut. Also a similar flat sliding bolt is used to lock forends in place that operate on the Anson & Deeley principle. Westley Richards also have a top bolt or, perhaps more correctly, a draw bolt.

Bolt run-up/Run-up The curved rear of the lump (front lump on a double bite gun) that the locking bolt will ride against until engaging the bite.

Bore The inside of the barrel also used to denote the size, e.g. twelve-bore, although the older term still used in the USA is gauge. May also be used to describe the operation of barrel boring, i.e. to bore a barrel.

Bottom plate/Cover plate The plate fixed to the bottom of a boxlock action to cover the mechanism. A modern term is frame plate.

Box The projection on a trigger plate to which the triggers are hinged and the breech pin screws into. Occasionally referred to as the turret.

Boxlock A type of gun where the lock mechanism is enclosed within the action.

Breech In London the end of the barrel that contains the cartridge. In Birmingham and most of the rest of the world it would be the chamber. Sometimes they are used together such as having a cartridge 'chambered in the breech'.

Breech pin/Action pin The oldest and most traditional term is undoubtedly breech pin and is the large pin (or countersunk head screw to a non-gunsmith) that goes through the top strap just behind the breech and screws into the box of the trigger plate. Alternative modern terms are main pin, body pin or frame screw.

Bridle In a sidelock or hammer gun the bridle secures the component parts of the lock in place with the retaining pins and hammer pivots passing through it. A ring bridle is where it is cut away in a skeleton form like a loop. A pierced bridle is cut away with a decorative shape.

Bridle pins The small screws that hold the bridle, sear etc. onto the lock plate.

Butt pad/Recoil pad A term used for comparatively soft rubber or Sorbethene pads fitted to the butt to help reduce, or at least give the impression of reducing, recoil. A term now used to encompass any type of 'soft' pad.

Butt plate/Heel plate Manufactured from either plastic, steel, horn, Vulcanite, Ebonite or even contrasting wood, and fitted to the butt to provide protection to the end of the stock and a slip-free fit to the user's shoulder.

Cam A projection either fitted to or part of the action bar knuckle or pivoted in the knuckle of the forend iron that pushes (or cams) open the extractor(s) when a gun is opened. Occasionally referred to as the extractor cam or extractor lever and on some imported guns as the ejector cam even when operating a non-ejector mechanism.

Cast The measurement of the amount the end (or butt) of the stock is offset to one side of the centreline of the barrel(s) to accommodate right- or left-handed shooters. Cast off – for a right-handed user – is when the stock is to the right viewed from above, cast on for a left-handed user.

Chamber The breech end of a barrel reamed out to accept a cartridge. *See also* Breech.

Choke The restriction just prior to the muzzle end of a barrel most often formed from a taper section leading to a parallel section smaller than the main bore and continuing to the muzzle. This has the effect of concentrating the shot pattern and increasing the distance at which clean kills can be made. Recessed choke is where a parallel bore is opened out for a few inches just before the muzzle, which remains at its original size. It is a modification employed by some present day competitive muzzle loader shooters to tighten the pattern of guns normally made without any choke. It is normal to assume that side-by-sides have most choke in the left barrel, over-and-unders (without screw-in chokes) in the top barrel. In the early days of choked breech loaders it was not particularly unusual for side-by-sides to have the most choke in the right barrel. This makes a lot of sense for driven game where the first shot is likely to be taken while the bird is furthest away. Sometimes these are referred to as grouse guns.

Chopper lump A method of barrel manufacture where the barrel tubes have flat extensions at right angles at the breech ends. They then resemble a simple form of axe commonly referred to as a 'chopper'. When brazed together as part of the barrel assembly these flat extensions form the basis of the barrel lumps, hence the term 'chopper lump'.

Cocking dogs The arm in a boxlock mechanism that cocks the hammers when the gun is opened. An alternative is limb(s) and one more likely to be encountered in the London trade. The similar component in a sidelock is the lifter.

Cocking indicators These are still favoured in some European countries in the form of small peg-like buttons that protrude through the top strap to indicate when a hammer is cocked. On firing the indicator drops flush. In the transition period from hammer guns, which can be seen to be cocked to those with wholly internal mechanisms, there were various methods devised to show a gun was cocked.

Cross bolt A type of locking device that bolts sideways through an extension on the breech end of the barrels. Often referred to as a Greener cross bolt after the original inventor, there are nonetheless a number of variations. With some the cross bolt protrudes when the top lever is pushed to the fully open position. On others the cross bolt is hidden behind a cap and does not show out of the side of the action. Normally of round section it is possible to encounter, usually on European guns, those that are of rectangular section. Over-and-unders like Merkel and the AYA Yeoman are fitted with large flat cross bolts that engage on lugs either side of the top barrel and are the sole (and quite efficient) method of locking the gun shut.

Cross-over stock This is the most common description for a stock which, when viewed from above, is bent in the form of a shallow 's'. They are made in this way to accommodate a shooter's sight deficiency. Especially in sidelock form where the top strap, trigger plate and lock plates may be curved, it is one of the highest expressions of the gunmaker's art and an older generation of shooters, not bound by the shackles of political correctness, often still call them 'cross-eyed' stocks. A less extreme form or 'half cross-over' is usually referred to as a central vision stock.

Cross pin The large pin fitted near the front of the action bar which provides a pivot for the hook. Usually a separate pin but sometimes machined as part of the action. The cross pin is also referred to as the pivot pin.

Damascus   Used to describe the type of barrel produced by forging together iron and steel which, when browned, shows a distinctive pattern. There are various grades from a coarse, rather broken and blotchy pattern to the most fine and intricate work made by twisting and forging thin wire together. It is believed that the last Damascus tubes made in the UK were at W. W. Greener's factory in 1912. Also referred to as Damascus twist or, in its plainest forms, simply as twist.

Disc set strikers   Discs screwed into the face of the standing breech that retain the strikers or firing pins. P. V. Kavanagh took this a stage further by having the disc the full diameter of the head of the cartridge to provide a limited amount of adjustable head space. A modern term for disc set strikers is bushed firing pins.

Double trigger   Two triggers on a double-barrelled gun which are normally made so that in side-by-side format the front trigger operates the right hand barrel, the back trigger operates the left. With two triggers on an over-and-under the front normally fires the bottom barrel, the rear fires the top. Nothing, however, can be taken for granted. I have come across an early example of a Joseph Lang hammer gun where the front trigger operated the left barrel and the rear the right. Probably the rarest trigger arrangement is three barrelled guns with three triggers.

Drag   If a trigger has a long pull to disengage the sear it is termed drag or trigger drag.

Draw/Circle   The curved front of a rear lump intended to pull the barrel towards the face of the action.

Drop points/Tear drops   Fancy carving at the end of a stock panel – some rather plain, others in a stylized leaf form. Some old hands still refer to them as bottle points.

Ejector kicker(s)   The spring operated device which strikes against the extractor to eject a spent cartridge. On over-and-unders and some modern designed side-by-sides the same part is called the ejector hammer.

Escutcheon   A shield-shape usually of silver or sometimes gold and occasionally brass or silver-plated brass set into the stock for the owner's initials or coat of arms. For many years fashion has dictated this is set into the underside of the stock but in muzzle loading days where there was no long top strap it was not uncommon for it to be fixed in the top of the stock in front of the hand. Escutcheon is also used to describe a fixing for a pin such as found on a single-barrel sidelock opposite to the lock.

Extractor(s)   The device that engages a cartridge rim and pushes it clear of the breech when a gun is opened. Traditionally whether the gun is a non-ejector or has an ejector mechanism this part is still called the extractor. With a side-by-side ejector gun they are simply split extractors (two piece). With simple single-barrelled guns where the ejector works on opening if the gun has been fired or not it is normal to refer to it as an ejector. Manufacturers of modern over-and-unders almost always refer to the same part as an ejector whether the permanently sprung type, such as Beretta, or the type with separate sprung hammers, such as Browning. Even makers of modern-designed side-by-sides usually apply the rule that a non-ejector gun is fitted with an extractor, while an ejector gun has ejectors. There are oddities in the description of these parts with individual maker's terms such as extractor plate (non-ejector) and at least one instance of a one-piece extractor being referred to as an ejector.

Extractor bed/Extractor way   The recessed cut-away into which the extractor sits.

Extractor leg   The projection that extends at right angles from the extractor to locate through the extractor hole between the barrels and contact the cam and/or ejector mechanism.

Extractor peg   On side-by-sides where the leg is round, or a split round in the case of an ejector gun, the peg fitted near the top forms a location in the extractor bed to prevent the extractor twisting out of line when pushed clear of the bed. With over-and-unders where the leg is shaped to fit a dovetail slide this provides a positive location that does not require a peg.

Fence   A term originating in the days of percussion muzzle loaders and referring to the raised cup shape carved into a muzzle loader breech plug. This gave a certain amount of protection to the user's eyes as the percussion cap was, when struck by the hollow nosed hammer, almost enclosed. Early hammer breech loaders often duplicated these deep protective fences but later heavier breech designs omitted this except for those with a thin raised section cut for decorative purposes only, hence the term carved fences. Modern guns fall into one of two types: ball fence or cut fence, the former being a semi-spherical design, the latter a simpler, perhaps potentially stronger but less elegant 'D' shape.

Flintlock A form of ignition that has been in use for much longer than the breech loader or the percussion muzzle loader. The principle of the flintlock is where a flint is struck against a steel or frizzen to shower sparks into a small pan which ignites to then fire the main charge through a touch hole. The flintlock appeared in several variants but qualifies as two basic designs. The simpler and earlier design, the snaphaunce, has a cover over the pan that has to be manually opened, while with the later design the pan cover is attached to the frizzen. As the cock holding the flint strikes against the frizzen it snaps back to lift the cover and expose the priming charge of fine powder. Late good quality flintlocks were superb pieces of work with platinum-lined vents and shaped waterproof pans even with gold lining.

Forcing cone/Lead-in The tapered area immediately in front of the chamber. The forcing cone tapers from the chamber down to the bore of the barrel.

Forend The complete assembly of wood and steel that makes up the forend.

Forend bolt The spring locking bolt typical of the Anson & Deeley type that engages with the forend loop to hold the forend assembly in place. Also in earlier guns the plain flat headed bolt that fits crosswise through the forend and loop.

Forend iron The steel component of the forend assembly with the knuckle at one end that fits against the action and the forward projecting stale or steel to which the wood is fixed. With the terms 'steel' in London and 'stale' in Birmingham it is interesting to consider whether this term came about due to a matter of pronunciation. For example, in 'Brummie' dialect, 'wheel' is pronounced 'whale', likewise 'meat' is 'mate', so is it possible steel became stale? This presupposes, of course, that the name originated in London, something no self-respecting member of the Birmingham trade would ever be likely to admit to.

On some modern produced guns the forend iron is described as the forend hanger and there is also one further area of confusion – the top of the knuckle of the forend iron can be described as the toe and the small mating step at the end of the action flats above the knuckle of the action bar as the nib. However, I have come across these terms transposed to describe the opposite part. Once again this seems to be a matter of regional difference and may, like so many things, have occurred due to little more than misunderstanding. The forend iron in most instances also carries the ejector mechanism where fitted.

Forend loop The fixing under the barrel(s) that holds the forend in place. With muzzle loaders and early breech loaders this was a slotted projection through which a flat forend bolt was fitted crosswise. Later it became a form of catch but is still described as a loop.

Forend pipe The tube at the end of the forend iron stale (or steel) in the Anson & Deeley design which encloses the plunger and spring that connects to the bolt.

Forend tip A decorative method of protecting the front edge of the forend wood, often a piece of inlet horn or Ebonite. On forends with a pipe it is made from steel and forms a collar around at the end of the pipe, very much like an abbreviated form of those used on muzzle loaders through which the ramrod fits and generally called a tail pipe.

Forend wood The part most gun owners recognise as the forend although the wood is only part of the forend assembly. British and European side-by-side guns traditionally adopted a small slim shape often called splinter forends in a variety of subtly elegant styles. Americans favour a larger, longer style that wraps around the side of the barrels and have earned the description beaver tail from its shape when viewed from the underside. On Continental and American over-and-unders it is more often described as forearm wood.

Furniture A general term for all the small parts and fittings such as trigger guard, forend tip etc.

Gape The amount a break-open gun has to open at the breech to allow loading. For example, a gun with superimposed barrels has to drop the barrels further (through a steeper angle) to open far enough for the bottom barrel to clear the breech face than a single or side-by-side. Therefore it is described as having a wider gape.

Hammer/Tumbler That part which hits the striker or firing pin to fire the cartridge. The similar part on a flintlock that holds the flint is the cock. External hammers on percussion guns were originally referred to as cocks but certainly by the time of the breech loader became hammers, although we still 'cock the hammers'. The external hammer is connected to a tumbler which is attached to the mainspring and cut with bents for the sear to engage. With internal hammers in sidelocks they become part of the tumbler and may be referred to as either hammer or tumbler.

Hand The slim section of a straight hand stock behind the action and the part which is held in the hand. Occasionally called the wrist or small of the stock, the chequered area is the grip. When this part of the stock is shaped in the form of a curve with a defined end it is either a bag grip (half pistol) with a rounded end or pistol grip (sharp curve with a flat end) seemingly regardless of the addition or otherwise of chequering. Another variation is the swan neck stock which gently curves down at the hand then flows around into the butt without any kind of grip end.

Hand pin/Long-hand pin The pin that passes through the trigger plate tang and the hand of the stock and screws into the top strap. A modern term is rear frame screw.

Headspace Effectively the distance between the base of the cartridge when chambered and the face of the breech. On new guns headspace is strictly toleranced. With wear, if it becomes excessive recoil increases and in extreme circumstances protruding primers may be visible.

Hinge discs/Trunnions A method of providing a two piece pivot location for some over-and-under shotguns.

Horns The extensions on the head of a stock, usually a sidelock, that engage into the back of the action as a form of strengthening.

Intercepting sear A safety sear which engages against the hammer to block it from hitting the striker or firing pin if the main sear should be faulty and slip out of engagement.

Keel rib The short rib behind the forend loop. When made as part of the forend loop it may still be referred to as the keel rib or sometimes keel plate or keelpiece.

Knuckle The knuckle has two parts in a conventional side-by-side or single: the convex part at the front of the action bar and the mating concave part of the forend iron.

Laidover rib A top rib that is laid the full length of the barrels on a side-by-side and regarded as the very best form of top rib.

Lifter The long cocking arm in a sidelock self cocking gun that links into the forend knuckle and cocks the hammer(s) when the gun is opened.

Lock pin This is the long (screwed) pin which passes through one lock plate of a sidelock and screws into the opposite lock plate or, in the case of a single-barrel gun, into an escutcheon or side plate. Quite rarely it might be referred to as the 'nail', a term more common in the flintlock era.

Lock plate The shaped plate to which the lock mechanism is fixed.

Mainspring/Hammer spring The large vee spring that operates the hammer. In Birmingham it is usually the mainspring and in London, the hammer spring.

Monoblock An economic method of manufacturing double-barrelled guns by using a shaped block at the breech end into which spigoted barrel tubes are soldered. Often regarded as a modern process, the original idea dates back to W. H. Monks' patent of 1881, albeit in that design the tubes were inserted from the breech end of the monoblock.

Muzzle The end of the barrel furthest from the breech.

Nipple On a muzzle loader, the cone-shaped device onto which a percussion cap is placed to be struck by the hammer and thereby ignite the main charge through a central hole drilled through the nipple. The similar device on a hammer breech loader that holds a striker (or firing pin) in place.

Nose end This is the shaped end of a good quality rib that fits down between the muzzle end of a pair of barrels.

Oval A silver, gold, brass or plated brass device set into the stock for the owner's initials, coat of arms or similar. An oval is actually nearly round and assumes the oval form of appearance when fitted around the shape of the stock.

Overdraft The amount of clearance between the lower edge of the barrels of a break-open gun when fully opened and the top of the breech face. This should be just sufficient to allow loading.

Panel(s) The shaped sides of a stock behind the action that in the case of a sidelock are cut out to accept the lock plate or for decoration on a boxlock.

Percussion gun A muzzle loader using a fulminate form of ignition, the ultimately successful design of which was the copper percussion cap. The term percussion gun differentiated quite clearly between that and a flintlock gun. If we wished to be pedantic, then breech loaders are also a form of percussion gun as striking the primer detonates the small charge that ignites the main propellant charge of powder.

Pin This is the traditional name for all the fixings which hold a gun together, whether plain dowel pins, cheese head or countersunk headed with full or part length threads. On many modern factory-made guns the threaded variety of pin is usually referred to as a screw which, in engineering terms, that is exactly what it is.

Proof marks Marks stamped on barrels and actions to show a gun has been subjected to and passed proving or proof testing. Those on the barrel(s) indicate bore size, proof load and chamber length.

Pull/Length of pull The dimensions from the trigger to the heel, middle and toe of the stock.

Rib(s) The joining pieces between the barrels on a side-by-side; the top rib, bottom rib and keel rib. An over-and-under has side ribs and a top rib which may be both raised and ventilated, i.e. with longitudinal gaps. Guns have been produced without ribs as the breech end is already fixed and only a joining piece is then needed at the muzzle.

Rim recess The recessed step at the breech end of a barrel into which the cartridge rim sits.

Root The area where the breech face of a side-by-side joins the flats. Originally of a sharp (so-called knife edge) design, it was fairly soon modified to a stronger radius form and is sometimes simply referred to as the radius where 'radius of the root' would be a bit of a mouthful. One American term is 'gusset' rather than radius.

Safety/Safety catch/Safety thumbpiece/Safety button All terms used to describe the safety operating device on either the tang of a gun, the side in the case of a Greener side safety, or in the trigger guard. A rare item on shotguns was the provision of a grip safety which had to be held depressed into the stock for the gun to be fired.

Safety slide/range In a gun with a simple automatic safety the slide or range is the plate or bar which is pushed back either by the top lever or the bolt to engage the trigger lock safety device. Three modern terms are safety bar, safety impeller or safety activator.

Sear The catch which engages with the hammer (or tumbler). When pushed out of engagement it releases the hammer to fire the gun.

Shield A decorative device carved into the top of the breech that aligns with the top rib and is often in the form of a simple shield. Not to be confused with the fancy shield-shape set in a stock for the owner's initials which is an escutcheon.

Side clips Small projections either side of the standing breech of a side-by-side which engage on the sides of the barrels. More popular on continental guns, on British guns they mainly appear on wildfowling guns intended for heavy loads.

Side plates Dummy plates fitted to boxlocks to give the appearance and styling of a sidelock.

Sidelock A gun made with the lock or locks fitted to the side. Sidelocks may have external hammers or be of hammerless design, bar lock with the mainspring lying forward in the action or back action locks with the mainspring and lockwork all to the rear of the action.

Sights Sighting devices as one might find on a rifle are as such unusual on a shotgun, although a device marketed as 'Aimpoint' was moderately popular some years ago as a training device. With both eyes open this appeared to project a red blob of light in the area the shot would strike. While the foresight bead on many side-by-sides is so tiny as to be insignificant, some clay pigeon shooters use both large foresight beads and a smaller one fitted approximately halfway along the top rib. There are other foresight devices that look a little like a string of fairy lights which are probably more of a prop to a shooter's confidence than any real practical benefit.

Single trigger Literally a single trigger to operate both barrels of a double gun. In non-selector form it is used to fire a gun in a set sequence, i.e. right then left of a side-by-side or bottom/top of an over-and-under. With a selector mechanism as is common on over-and-unders and less so with side-by-sides, the sequence can be changed. Single triggers even appeared on such exotica as three-barrelled guns. While I understand a single trigger arrangement has been adapted to the four-barrelled form of gun it should not be confused with the early design of Lancaster, where a single trigger is connected to a rotating hammer that gives a long pull and really not suited to shotgun use. Occasionally there are double trigger guns where the front trigger can be used as a single trigger; in other words pulled twice to fire the two barrels in a set sequence, usually bottom/top, and the rear trigger is only needed if the bottom barrel is required to be fired first. Even variations on this idea have been tried where each trigger of a double trigger gun can be used to operate both barrels. Three-barrelled guns like the Drilling shotgun/rifle combination guns usually have one trigger: the front, made to fire the right hand shot barrel and rifle barrel but operated by a manual selector (which cunningly also flips up the rear sight), leaving the rear trigger to operate the other shot barrel.

Spindle The semi-rotating cylindrical link between the top lever and locking bolt.

Stock The wood between the action and where it mounts against the user's shoulder. The stock comprises of the head, where it fits against the action; the hand, being the slim oval section where it is held; the grip, being the chequered area; the comb, which is the line across the top of the stock to the butt. At the butt end of the stock there is from the top downwards, the bump or heel, the middle and the toe, and the side of the stock that is closest to the user is the face.

Stock bolt A method of securing the stock on most over-and-unders and some side-by-sides, mainly of modern design. The stock bolt fits in the stock and screws into a threaded hole at the rear of the action or tangs, pulling the head of the stock into place around the tang and against the rear of the action.

Striker(s) The piece which strikes the primer of the cartridge to initiate ignition – what the rest of the world calls a firing pin. On mainly British boxlocks the striker is formed as part of the hammer and is called a dogtooth striker, due to its shape.

Swivel The link in a sidelock between the tumbler and the mainspring, and dimensionally one of the most important parts in the lock. Occasionally referred to as the stirrup.

Tang screws/Tang pins Those wood screw-type screws which secure the tang to the stock, sometimes called trigger guard screws.

Top extension/Top barrel extension An extension from the breech face of the barrels, which engages into a cut-out in the top of the standing breech. Some form a third bite and can be made in such a way as not to be visible when the gun is shut. Others, such as the extension for a Greener crossbolt or particularly the doll's head extension (so called due to its shape), remain visible and are engraved to match any decoration on the breech.

Top lever Opening lever mounted on the top of an action behind the breech. At least one modern manufacturer refers to this as the top snap, a term in use in the nineteenth century, that was certainly applied to the Westley Richards with the single bite in the doll's head extension.

Top strap The extension behind the action of a traditional side-by-side through which the breech pin is fitted and on most, but not, all guns the long hand pin screws into. Sometimes referred to as the tang, it seems generally accepted that only that part which carries the safety button or thumb piece – especially if stepped in to form a smaller width – might be referred to as the tang.

Trigger The trigger(s) in a traditional gun is formed from the curved finger piece and the flat blade which operates against the sear. To add a bit of confusion the finger piece on a rifle is often referred to as the blade, a term which seems to be creeping into use on some shotguns.

Trigger guard A trigger guard may be of the short variety without a tang or, more traditionally, with a long tang either straight or curved (to fit a curved grip). Particularly in the straight form the tang is sometimes described as the tail. The bow wraps around the trigger(s) and may be made with a rolled edge – a fancy rounded edge – on the side the trigger finger lies and in this form is known as single bead; double bead has a rounded edge both sides.

Trigger plate The plate to which the triggers are fitted.

Trigger plate action An action where the lockwork, triggers, hammers, springs etc. are mounted on the trigger plate.

Try gun A test gun with an adjustable stock so that length, cast and bend or drop can be altered to suit the user and so determine the actual stock dimensions required for a customer's own gun.

Vent(s) A device to bleed gas away from a shooter's face in the event of a ruptured or failed primer. A fairly sophisticated arrangement is where hollow vent pins engage with discs screwed into the breech face. A much simpler method is a shallow annular ring machined into the breech face around the striker or firing pin hole with a joining cut-out leading to the side of the breech face.

Section Two

Pump actions and Semi-automatics – Parts/Description

Many parts of these guns still use traditional names. Obviously a barrel is still just that and similarly a sear is also still a sear, although quite different in design to that, say, in a side-by-side boxlock. Pins, of which in some designs there seem to be a multiplicity, are usually plain dowel pins, some with annular cut-outs to engage with a spring retainer, others are the self-sprung roll pins. It should be remembered that these are engineered guns so screws and bolts are (usually) exactly as would be described by an engineer except in instances where the moving breech might be described as the breech bolt and, just to add a little confusion, the breech block. So, while there is some standardization of the naming of parts there are still different descriptive names for parts that carry out the same function.

The parts listed below in this section are those which have functions or names peculiar to these types of gun.

Action bar/Cocking rod The bar which connects with either an action sleeve (pump or slide action) or piston assembly (semi-auto) and operates the breech bolt or block.

Action slide sleeve screw cap A screwed ring to hold the forend wood in place on the action sleeve assembly.

Action slide/Slide The assembly made up of a sleeve that fits over the magazine and inside the forend and one or two long bars connected to the breech bolt or block.

Bolt latch The latch that holds the breech open after the last cartridge from the magazine has been fired or when the breech is opened for loading.

Breech bolt/Breech block The moving breech piece which locks shut on closing and carries the firing pin, extractors and locking block.

Butt/Buttstock The wood or sometimes a synthetic material between the back of the receiver and the butt plate.

Carrier/Elevator The shaped plate which lifts a cartridge from the magazine to the breech.

Cocking rod support Basically the same component as an action sleeve.

Ejector On these guns it may be a projecting lug or something as simple as a cut-out in the barrel extension that catches one side of the cartridge rim throwing it through the ejector port.

Ejector port Cut-out in the receiver to enable a cartridge to be loaded direct into the breech and for cases to be ejected out of with the gun in use.

Firing pin On these type of guns always a firing pin, not a striker.

Forearm/Forend The forend wood that fits around the slide of a pump action or magazine of a semi-auto.

Forearm cap/Forend cap Screw cap which holds the magazine spring, magazine follower and magazine spring retainer in place. On many semi-autos it also secures the forearm/forend.

Magazine cut-off A device that can be used to block the cartridge feed from the magazine. This still enables the gun to be used as a single shot while keeping a full magazine for when rapid fire becomes desirable.

Magazine follower A simple bucket-shaped sleeve which fits between the magazine spring and the cartridges. It is stepped to prevent it leaving the tubular magazine after the last cartridge has been fed to the breech.

Magazine latch A sprung arm designed to retain cartridges in the magazine and only release one at a time as the breech bolt/block moves rearwards and the carrier moves upwards to hold another cartridge in place for loading.

Magazine spring A long helical spring used to push cartridges out of the magazine as the magazine latch permits.

Operating handle The projecting lug from a semi-auto breech bolt/block used to open the breech and/or cock the mechanism.

Piston/Piston assembly A part peculiar to gas operated semi-autos. The piston assembly is pushed rearwards by gasses bled from a block under the barrel. Linked to the action bar or cocking rod it automatically operates the mechanism to open the breech.

Receiver The hollow steel or alloy receiver which holds the mechanism and to which the butt and barrel are fitted. While it is normal to refer to ancillary items as action parts, the main housing is almost always described as the receiver.

Recoil spring Recoil springs are peculiar to semi-autos as a pump action does not need one due to the entirely manual operation. Gas operated semi-autos carry the recoil spring in the butt housed in an action recoil spring tube while recoil operated guns have a large diameter spring fitted around the magazine tube.

Section Three

Processes

Barrel lining Fitting a full length liner tube up a barrel as a method of repair.

Barrel sleeving A means of providing new barrel tubes by spigoting them into a truncated section of the original barrels and tinning (soldering) in place.

Bending Usually applied to bending a stock to produce a different cast, bend or drop. Probably the most common method is using hot linseed oil; steaming has been popular but there is a growing use of domestic spotlights to focus heat in a small area of the hand of the stock.

Blacking Producing a black chemical finish on barrels and furniture, also sometimes actions as often found on BSA double-barrelled shotguns. Blacking can be divided into hot water or express blacking, cold blacking and caustic blacking. Caustic blacking is not suitable for double-barrels as even with silver brazed ribs there is always a very good chance the hot salts will find a way into the cavity between the barrels and leech out later, often for many months, even years.

Blacking down The process of fitting gun parts together using smoke blacking to highlight the contact areas.

Blazing off Burning away oil to temper a spring.

Bluing A term usually reserved for charcoal or heat bluing of parts, literally giving a blued finish.

Boring The method of opening out a barrel bore either to finish or refinish or to remove choke. The traditional means of achieving this was using a spill borer.

Browning At one time a name used to encompass both the barrel finishes of blacking and browning. Now used to describe a browned finish, particularly on Damascus barrels.

Case hardening Forming a hard carbon rich layer on the outside of component parts to give good wear characteristics. May be achieved with proprietary compounds.

Chambering Cutting a chamber in the breech end of a barrel with a chamber reamer to accept (chamber) a cartridge.

Colour case hardening A process to both harden and enhance the appearance of steel parts. Carried out in the traditional way of packing the parts in a steel container with a mixture of carbon rich materials, mainly bone meal. Heated for several hours (soaking being the rather odd term), the parts are plunged direct from the steel container into water. This process both hardens and gives the characteristically subtle yellow, blues and browns on many actions.

Dent raising The raising of dents from a barrel using an hydraulic or mechanical slipper dent raiser and a specially shaped hammer.

Draw boring Lapping a barrel bore (*see* lapping).

Heading up/Heading on Fitting a stock to the rear of an action.

Honing A method of bore finishing or refinishing using a mechanical carrier and abrasive stone(s).

Lead lapping A method of bore finishing or refinishing using a soft lead lap and abrasive powder mixed in oil.

Oil finish A method of stock finishing with oils, usually with a linseed oil base that gives a particularly attractive, durable and comparatively easy to repair finish. 'London finish' or 'best London finish' includes a shellac based finish for the final coats.

Oil hardening A process where steel may be altered to its working hardness by heating and quenching in oil or used simply to harden steel prior to tempering.

Rimming Cutting the recess in the end of a barrel

to accept the cartridge rim. Some chamber reamers incorporate a section for the rim but most often for shotguns rim cutters are a separate device.

Silver brazing Commonly called silver soldering and used to hold dovetail lump barrel assemblies together, also the ribs on some continental guns. May be obtained in a variety of alloys that give a wide range of working temperatures.

Smoke blacking Also sometimes simply called blacking, it might seem to cause confusion to an outsider with the chemical finishing process. Similarly the smoke lamp may be described as a smoker, blacking lamp or blacker, the latter of course is also someone who carries out the finishing process of blacking. A good guide is the context in which the word is being used.

Striking up The finishing process on steel parts prior to final polishing. A term most often used to describe refinishing of barrels before blacking. In this state parts are known as being 'in the white'.

Sweating on Soldering an item in place. When heated with a torch condensation forms on the surrounding steel, probably the origin of the term 'sweating' on.

Tempering After hardening, springs in particular are reheated to a lower temperature which renders them springy (after just hardening they would snap). Also used on some components where the required characteristic is hard and tough rather than hard and brittle. Sometimes called 'letting down', this term is more normally reserved for annealing (or softening) of previously hardened parts so they can be reworked or repaired.

Tinning Preparing the parts for a soldered joint by applying a very thin coating of solder to each part prior to assembly.

Index